2009 INTERNATIONAL BUILDING CODE NEED TO KNOW

About the Author

R. Dodge Woodson has written more than 100 non-fiction books. He is a licensed general contractor and a licensed master plumber who has built as many as 60 single-family homes a year. Woodson is a well-known remodeling contractor, plumbing contractor, and business owner. Additionally, Woodson is accredited as an expert witness and serves as a consultant on construction and plumbing litigations. He lives in Brunswick, Maine.

2009 INTERNATIONAL BUILDING CODE NEED TO KNOW

The 20% of the Code You Need 80% of the Time

R. Dodge Woodson

New York Chicago San Francisco Lisbon London Madrid
Mexico City Milan New Delhi San Juan Seoul
Singapore Sydney Toronto

The McGraw·Hill Companies

Cataloging-in-Publication Data is on file with the Library of Congress

2009 International Building Code Need to Know

2 3 4 5 6 7 8 9 0 WFR/WFR 0 1 4 3 2 1 0

ISBN 978- 0-07-159257-4
MHID 0-07-159257-1

Sponsoring Editor
Joy Bramble Oehlkers

Acquisitions Coordinator
Michael Mulcahy

Editorial Supervisor
David E. Fogarty

Project Manager
Jacquie Wallace, Lone Wolf Enterprises, Ltd.

Copy Editor
Jacquie Wallace, Lone Wolf Enterprises, Ltd.

Proofreader
Roger Woodson, Lone Wolf Enterprises, Ltd.

Indexer
Leona Woodson, Lone Wolf Enterprises, Ltd.

Production Supervisor
Pamela A. Pelton

Composition
Lone Wolf Enterprises, Ltd.

Art Director, Cover
Jeff Weeks

Dedication

This book is dedicated to my daughter, Afton.
She graduated magna cum laude this year with
two majors and a minor at the age of 20.
Afton begins work on her master's degree later this summer.
I have many accomplishments in my life,
but my children Afton and Adam are by far
my two greatest accomplishments.

Contents

Preface xiii
Acknowledgments xv

CHAPTER 1
Use and Occupancy Classifications 1
Group A-1 .... *2*
Group A-2 .... *2*
Group A-3 .... *3*
Group A-4 .... *3*
Group A-5 .... *4*
Business Group B *4*
Group E *5*
Group F *6*
Group F-1 .... *6*
Group F-2 .... *8*
Group H *8*
Group S-1 .... *10*
Institutional Group I *10*
I-3 Facilities *11*
Group I-4 .... *13*
Group M *13*
Group R *13*
Group S *15*
Group U *16*

CHAPTER 2
Special Detailed Requirements Based on Use and Occupancy 17
Covered Mall Buildings *18*

Fire-Resistance Separations .21
Children's Playgrounds .22
Security .24
Plastic Signs .24
High-Rise Buildings .25
Underground Buildings .27
Group I-2 Medical Facilities .29
I-3 Different Occupancies .32
Power-Operated Sliding Doors .34
Vertical Openings .35
Stages and Platforms .37
Stage Doors .38
Combustible Materials .40
Roof Vents .40
Platform Construction .41
Dressing Rooms .41
Automatic Sprinkler Systems .42
Aircraft-Related Occupancies .42
Hangers .43
Combustible Storage .46
Hazardous Materials .46
Storage .50
Hazardous Materials with Rankings of 3 or 4 .50
Occupancy Groups H-1, H-2, H-3, H-4, and H-5 .50
Conveyors .53
Mixed Occupancies .53
Group H-5 .55
Gas Detection .59
Emergency Power .60
Exhaust Ventilation .61
Occupancy Groups I-1, R-1, R-2, and R-3 .61

CHAPTER 3

General Building Heights and Areas .63
Definitions .63
General Height and Area Limitations .64
Height .65
Mezzanines .65
Equipment Platforms .67
Area Modifications .68

Unlimited Area Buildings *70*
Mixed Use and Occupancy *73*

CHAPTER 4
Fire-Resistance-Rated Construction **75**
Fire-Resistance Rating and Fire Tests *75*
Exterior Walls *77*
Outside Walls *78*
Parapet Construction *80*
Fire Walls *81*
Fire Barriers *84*
Smoke Barriers *85*
Smoke Partitions *85*
Horizontal Assemblies *86*
Ceiling Panels 86
Penetrations *87*
Fire-Resistance Joint Systems *89*
Fire-Resistance Rating of Structural Members *90*
Exterior Structural Members 92
Opening Protectives *93*
Doors 93
Fire Dampers *97*
Smoke Damper 98
Draft Stopping 101
Loose-Fill Insulation 103

CHAPTER 5
Fire Protection Systems **105**
Definitions *106*
Automatic Sprinkler Systems *107*
Group A 107
Group A-3 and Group A-4 108
Group A-5 108
Group E 109
Group F-1 109
Group H 109
Group I 110
Group S-1 111
Monitoring and Alarms *111*
Alternative Systems *113*
Commercial Cooking *114*

Standpipe Systems 116
Class I 116
Class II and Class III Standpipe Systems 119
Alarm and Detection Systems 120
Smoke Control 121

CHAPTER 6

Means of Egress 125
General Means of Exits 125
Occupant Load 128
Means of Egress Illumination 129
Accessible Means of Egress 130
Doors, Gates, and Turnstiles 134
Floors and Landings 138
Door Handles 139
Stairways 140
Ramps 144
Exit Signs 145
Handrails 146
Guards 147
Exit Access 148
Aisles 150
Exit and Exit Access Doorways 150
Floor Number Signs 153
Smokeproof Enclosures 153
Horizontal Exits 154
Exterior Exit Ramps and Stairways 155
Emergency Escape and Rescue 156

CHAPTER 7

Accessibility 159
Scoping Requirements 159
Accessible Routes 161
Accessible Entrances 163
Parking and Passenger Loading Facilities 165
Dwelling Units and Sleeping Units 166
Special Occupancies 169
Other Features and Facilities 170
Platform Lifts 171
Signage 171

CHAPTER 8

Exterior Walls 173

Performance Requirements 173

Materials 175

Flashing 176

Veneers 177

Window Sills 181

CHAPTER 9

Roof Assemblies and Rooftop Structures 183

Performance Requirements 185

Fire Classification 186

Materials 187

Requirements for Roof Coverings 187

Roof Insulation 190

Rooftop Structures 191

Reroofing 192

CHAPTER 10

Structural Design 195

Construction Documents 196

General Design Requirements 198

Load Combinations 200

Dead Loads 202

Live Loads 202

Snow and Wind Loads 205

Soil Lateral Loads 207

Rain Loads 208

Flood Loads 208

Earthquake Loads 210

CHAPTER 11

Soils and Foundations 213

Foundation and Soils Investigations 214

Excavation, Grading, and Fill 216

Allowable Load-Bearing Values of Soils 219

Footings and Foundations 219

Pier and Pile Foundations 223

Driven Pile Foundations 228

Micropiles 229

Pier Foundations 230

CHAPTER 12
Wood **233**
Minimum Standards and Quality *234*
Fiberboard 235
Trusses 235
Fire-Retardant-Treated Lumber 236
General Construction Requirements *236*
Decay and Termites 238
General Design Requirements *238*
Allowable Stress Design *239*
Conventional Light-Frame Construction *240*
Floor Joists 240
Braced Wall Panels 241
Purlins 241
Engineered Wood Products 241

CHAPTER 13
Plastic **245**
Foam Plastic Insulation *245*
Interior Finish and Trim *249*
Plastic Veneer *250*
Light-Transmitting Plastics *250*
Light-Transmitting Plastic Wall Panels *251*
Light-Transmitting Plastic Glazing *252*
Light-Transmitting Plastic Roof Panels *253*
Light-Transmitting Plastic Skylight Glazing *254*
Light-Transmitting Plastic Interior Signs *254*

CHAPTER 14
Encroachments Into the Public Right-of-Way **255**

Appendix—Reference Material 257

Index 283

Preface

This book is your guide to getting your code questions answered with less stress. It is your ticket to simplifying the cryptic code that you work with on a daily basis. The code is large and can be complicated. I've translated the code into easy-to-use terms for people in the field. See the code as a real-world guide instead of some foreign language that only some people can begin to sift through. This is your chance.

How important is understanding the code? It is essential for professionals in the trade. Building without code compliance is an expensive mistake. Many good builders and contractors mean well and still stray from the code. This is often due to the code being difficult to understand and follow. Not anymore. This book will walk you through the code requirements chapter by chapter.

Laid out in the same basic sequence as the code book, this book can be used in conjunction with your code book. While this book is not a replacement for your code book, it is a strong enhancement to it. Use the books together as resources to make your job or your business more productive.

Thumb through the pages here. Notice the tip boxes. You will find that some key components of the code are highlighted in the tip boxes. Go ahead, take a peek. While you are at it, pick a few topics that you are either familiar with or confused about, and look them up. See for yourself how easy this book makes it to put your mind at ease on code issues.

You might find another book that attempts to do what this one does, but if you are looking for one guide to the code that you can trust, this is it.

Acknowledgments

I would like to thank the International Code Council, Inc. for its cooperation and permission to reprint certain illustrations and tables.

2009 INTERNATIONAL BUILDING CODE NEED TO KNOW

CHAPTER

1

Use and Occupancy Classifications

Structures or portions of structures need to be classified in regards to the occupancy of one or more groups. A room or space that is intended to be occupied at different times for different purposes must comply with all of the requirements that are applicable to each of those purposes. These structures that you have identified as having multiple uses, or occupancies, must comply with the code book. You may have questions about a structure that is proposed for a purpose that is not found in the code book. In this event, each structure must be classified in the group that the occupancy most nearly resembles, according to the fire safety and hazard involved. The 2009 International Code Book lists ten categories or groups that a structure or portion of a structure must be classified as. You will find clarification on these classifications as we go through this chapter.

- Assembly: Groups A-1, A-2, A-3, A-4, and A-5
- Business: Group B
- Educational: Group E
- Factory and Industrial: Groups F-1 and F-2

- High Hazard: Groups H-1, H-2, H-3, H-4, and H-5
- Institutional: Groups I-1, I-2, I-3, and I-4
- Mercantile: Group M
- Residential: Groups R-1, R-2, R-3, and R-4
- Storage: Groups S-1 and S-2
- Utility and Miscellaneous: Group U.

Group A occupancy includes, among other things, using a building or structure or even a portion of building or structure for purposes such as a civic, social, or religious functions; recreation, the consumption of food or drink; or for those waiting for transportation, such as a bus or taxi. There are exceptions in the classification of Assembly Group A:

- A building used for assembly purposes that has occupancy of less than 50 persons must be classified as Group B occupancy.
- A room or space used for assembly purposes that is less than 750 square feet in area and is also accessible to another occupancy must be classified as a Group B occupancy.

GROUP A-1

A-1 Group assemblies provide fixed seating and are intended for the use of producing and viewing performing arts and movies. This group includes, but is not limited to, the following:

- Motion picture theaters
- Symphony and concert halls
- Television and radio studios that include live audiences
- Theaters.

GROUP A-2

A-2 Group assemblies are intended for the consumption of food and/or drink and include, but are not limited to, the following:

- Banquet halls
- Night clubs

- Restaurants
- Taverns and bars.

GROUP A-3

A-3 Group assemblies are intended for worship, recreation, or amusement or other assembly uses but are not classified elsewhere in Group A, and include:

- Amusement arcades
- Art galleries
- Bowling alleys
- Places of religious worship
- Community halls
- Courtrooms
- Dance halls (without food or drink)
- Exhibition halls
- Funeral parlors
- Gymnasiums (without spectator seating)
- Indoor swimming pools (without spectator seating)
- Indoor tennis courts (without spectator seating)
- Lecture halls
- Libraries
- Museums
- Waiting areas in transportation terminals
- Pool and billiard parlors.

GROUP A-4

A-4 Group assemblies are those with the intended use for viewing indoor sporting events and activities that allow for spectator viewing. These include, but are not limited to, the following:

- Arenas

- Skating rinks
- Swimming pools
- Tennis courts.

GROUP A-5

A-5 Group assembly uses are intended for participation in or viewing outdoor activities. Some of these include, but are not limited to, the following:

- Amusement part structures
- Bleachers
- Grandstands
- Stadiums.

Many buildings may have multiple uses of occupancies and are referred to as mixed occupancies. In cases with mixed occupancies, they are required to meet the codes for those specific areas. Simply put, this means that occupancy of one group may also qualify for another. When this occurs, the stricter code is usually enforced by code officials.

BUSINESS GROUP B

Business Group B occupancy is the use of a building or structure for office, professional, or service-type transactions. Even if a person is using partial space of a building for an office, it would fall into this group. The storing of records and/or accounts is included in this as well. Below is a list of business occupancies for group B. Keep in mind, there may be more that qualify that are not listed here.

- Airport traffic control towers
- Animal hospitals, kennels, and pounds
- Banks
- Barber and beauty shops
- Car washes

- Civic administration offices
- Clinic-outpatient
- Dry cleaning and laundries: pick-up and delivery stations and self-service
- Educational occupancies for students above grade 12
- Electronic data processing
- Laboratories: testing and research
- Motor vehicle showrooms
- Post offices
- Print shops
- Professional services (architects, attorneys, dentists, physicians, engineers, etc.)
- Radio and television stations
- Telephone exchanges
- Training and skill development not within a school or academic program.

GROUP E

Educational Group E occupancy is the use of a building or structure, and again, this can even be a portion of a structure, for educational purposes through grade 12. However, there must be six or more people at any one time for this structure to be classified as this group. You should note that religious education rooms and religious auditoriums which are accessory to places of worship fall under A-3 occupancies.

CODE UPDATE

Mercantile Group M applies to buildings where merchandise is displayed and the sale of goods occurs.

Day care and child care or personal care services for a group of five children who are over the age of two and a half is also placed in the Group E occupancy.

GROUP F

The following list describes Group F, Factory Industrial Group that do not qualify as a Group H hazard or Group S storage.

- Assembling
- Disassembling
- Fabricating
- Finishing
- Manufacturing
- Packaging
- Repairing
- Processing.

GROUP F-1

Group F-1 describes the moderate-hazard occupancy of the factory industry. F-2 (low factory hazards) occupancies must never be included in this list and are in a league of their own which will be described in the next section. The following are included in the F-1, Moderate-hazard Occupancy Group:

- Aircraft
- Appliances
- Athletic equipment
- Automobiles and other motor vehicles
- Bakeries
- Beverages: over 12-percent alcohol content
- Bicycles
- Boats

- Brooms or brushes
- Business machines
- Cameras and photo equipment
- Canvas or similar fabric
- Carpets and rugs, this includes cleaning
- Clothing
- Construction and agricultural machinery
- Disinfectants
- Dry cleaning and dyeing
- Electric generation plants
- Electronics
- Engines, this includes rebuilding
- Food processing
- Furniture
- Hemp products
- Jute products
- Laundries
- Leather products
- Machinery
- Metals
- Millwork (doors and sash)
- Motion pictures and television filming (without spectators)
- Musical instruments
- Optical goods
- Paper mills and products
- Photographic film
- Plastic products
- Printing or publishing
- Recreational vehicles

- Refuse incineration
- Shoes
- Soaps and detergents
- Textiles
- Tobacco
- Trailers
- Upholstering
- Wood: distillation
- Woodworking (cabinet).

GROUP F-2

F-2, Low-hazard Occupational Industry, are uses that involve the fabrication of noncombustible materials. The following materials do not involve a significant fire hazard:

- Beverages: containing up to 12-percent alcohol
- Brick and masonry
- Ceramic products
- Foundries
- Glass products
- Gypsum
- Ice
- Metal products.

GROUP H

Group H is classified as a high-hazard group that involves the manufacturing, processing, generation, or storage of materials that have a physical or health hazard. Hazardous uses are classified in Groups H-1, H-2, H-3, H-4, and H-5. Keep in mind that these groups not only have to comply with the code, but also the International Fire Code. There are exceptions to Group H and the following are not to be classified as Group H, but must be classified in the occupancy that they most nearly resemble.

- Wholesale and retail sales and storage of flammable and combustible liquids in mercantile occupancies conforming to the International Fire Code
- Closed piping containing flammable and/or combustible liquids or gases that is used to operate machinery or equipment
- Cleaning establishments that use combustible liquid solvents having a flash point of 140°F (60°C) or higher in closed systems employing equipment listed by an approved testing agency, provided that this occupancy is separated from all areas of the building by one-hour fire barriers or one-hour horizontal assemblies, or both.
- Cleaning establishments that utilize a liquid solvent having a flash point at or above 200°F (93°C)
- Liquor stores and distributors without bulk storage
- Refrigeration systems
- The storage or utilization of materials for agricultural purposes on the premises
- Stationary batteries used for emergency power, provided that the batteries are provided with safety venting caps (note that ventilation is in accordance with the International Mechanical Code)
- Corrosives cannot and do not include personal or household products in their original packaging used for retail or commonly used building materials.

CODE UPDATE

An outpatient clinic is a building or a portion of a building that is used to provide medical care on less than a 24-hour basis to individuals who are not rendered incapable of self-preservation by the services provided.

GROUP S-1

All buildings and structures occupied for aerosol storage must be classified as Group S-1, and again, must meet the requirements of the International Fire Code.

The storage of black powder, smokeless propellant, and small arms primers in Groups M and R-3 and special industrial explosive devices in Groups B, F, M, and S must conform to the quantity limits and the International Fire Code.

INSTITUTIONAL GROUP I

Institutional Group I occupancy is the use of a building or structure in which people are cared for. This can also be a live-in supervised environment as well. People who qualify for this group have physical limitations due to health or age. Some may need medical treatment or other care. Institutional occupancies are classified as Group I-1, I-2, I-3, or I-4. Qualifications for Group I are housing of more than 16 people on a 24-hour basis, including those receiving personal care because of age, mental disability, or other reasons, but capable of responding to an emergency situation with physical assistance from staff. These residencies include, but may not be limited to, the following:

- Residential board and care facilities
- Assisted living facilities
- Halfway houses
- Group homes
- Congregate care facilities
- Social rehabilitation facilities
- Alcohol and drug centers
- Convalescent facilities.

If any of the facilities above have five or fewer residents, then the facility must be classified at a Group R-3 and must comply with the International Residential Code. If one of the facilities above has at least six, but not more than 16 persons, then it must be classified as Group R-4. Group I-2 occupancy includes buildings and structures that are used for medical, psychiatric, nursing, or custodial care on a 24-hour basis for more than five persons. Care must be available on a 24-hour basis and the persons are not capable of self-preservation. This group can include hospitals, nursing homes (to include intermediate and skilled care), mental hospitals, and detoxification facilities. A facility such as these that has five or fewer persons must be classified as Group R-3. One thing that you must remember is this: All facilities that provide 24-hour care and are considered to be live-in care, must comply with the International Residential Code. Please note that the above facilities do not include children under the age of 2 ½. A child care facility that provides care to five or more children in this age bracket need to be classified as Group I-2. There is an exception to this classification: A child-care facility that provides care to more than five, but no more than 100 children, and when the rooms in which the children are cared for are located on the same level as an exit and each of those rooms has a door directly to the outside shall be classified as Group E.

I-3 FACILITIES

I-3 facilities are occupied by persons who are generally incapable of self-preservation due to security measures that the occupants have no control over. This group may include:

- Prisons
- Jails

CODE UPDATE

High-hazard Group H occupancy includes, among others, the use of a building or structure, or a portion thereof, that involves the manufacturing, processing, generation or storage of materials that constitute a physical or health hazard in quantities in excess of those allowed in control areas.

- Reformatories
- Detention centers
- Correctional centers
- Prerelease centers.

Buildings that qualify for Group I-3 must be classified as one of five occupancy conditions. The first condition includes buildings in which free movement is allowed from sleeping areas and other spaces where access or occupancy is permitted to the outside of the building by means of an exit without having to be restrained. The code allows for a Condition I facility to be constructed as Group R. Condition II includes buildings which allow free movement from sleeping areas to an occupied smoke compartment to one or more other smoke compartments. Any outside exits are locked. Condition III allows free movement within smoke compartments to group activity spaces, but the means of exit are impeded by remote-control devices. Condition IV includes buildings in which free movement is restricted from an occupied space. Remote-controlled releases are used to allow movement from sleeping units, activity spaces, and other occupied areas within the smoke compartment to other smoke compartments. If your facility is classified with a Condition V then free movement is restricted from an occupied space. Staff control the door locks and a manual release is needed to permit movement from sleeping areas to other occupied areas within the smoke compartment.

The next section that I want to focus on is day care facilities. These buildings are of the Group I classification.

CODE UPDATE

A physical hazard is known as a chemical for which there is evidence that it is a combustible liquid, cryogenic fluid, explosive, flammable, organic peroxide, oxidizer, oxidizing gas, pyrophoric, unstable material, or water-reactive material.

GROUP I-4

Group I-4 is a building or structure that is used as a day care facility. These can be used or occupied by persons of any age who are receiving custodial care for less than 24 hours by individuals other than parents or guardians, blood relatives, marriage, or adoption. A facility such as above with five or fewer persons is classified as Group R-3 and must comply with the International Residential Code. Places of worship during religious functions are not included.

This group also includes adult care facilities. These provide accommodations for less than 24 hours and have five or more unrelated adults who need supervision and personal care. Take note that there is an exception to this group: A facility where occupants are capable of responding to an emergency situation without physical assistance from the staff is to be classified as Group A-3.

GROUP M

Group M is for Mercantile. Mercantile groups consist of buildings that display and sell merchandise. They include stocking of goods, wares, and merchandise for purchase from the public. These buildings can include:

- Department stores
- Drug stores
- Markets
- Motor fuel-dispensing facilities
- Retail or wholesale stores
- Sales rooms.

GROUP R

Group R is for residential groups that include the use of a building for sleeping purposes that are not classified as an Institutional Group I or when not regulated by the International Residential Code. R-1 residential occupancies include boarding houses, hotels, and motels that are transient in nature. R-2 residential occupancies are those that contain sleeping units or more than two living units. These are primarily permanent in nature and include:

- Apartment houses
- Boarding houses
- Convents
- Dormitories
- Fraternities and sororities
- Hotels
- Monasteries
- Motels
- Vacation timeshare properties.

R-3 residential occupancies are those that are permanent in nature and are not classified as R-1, R-2, R-4, or I. These include the following:

- Buildings that do not contain more than two living units
- Adult facilities that provide accommodations for five or fewer persons of any age for less than 24 hours
- Child care facilities that provide accommodations for five or fewer persons of any age for less than 24 hours
- Congregate living facilities with 16 or fewer persons.

R-4 is defined as residential occupancies that include buildings arranged for occupancy as residential care/assisted living facilities including more

Did You Know?

There are many groups and many qualifying information that you need to refer to. Always check with the codes to make sure that your facility has the correct qualifications or you may lose any special licensing that allows you to provide such care.

than five, but not more than 16 occupants, including staff. All R-4 occupancies must meet the requirements for construction as defined for Group R-3, except as otherwise provided in this book or the International Residential Code.

GROUP S

Group S is the storage group. This group will tell you what you may store in your building or structure that is not classified as a hazardous occupancy. Group S-1 is moderate-hazard storage and includes, but is not limited to aerosols of levels 2 and 3, baskets, boots and shoes, furs and leathers. Group S-2 includes buildings used for the storage of noncombustible materials such as products on wood pallets or in paper cartons with or without divisions and products such as knobs, handles, or film wrapping. Storage uses include, but are not limited to, storage of the following:

- Aircraft hangar
- Beverages containing up to 12-percent alcohol in metal, glass, or ceramic containers
- Chalk and crayons
- Dry cell batteries
- Empty cans
- Food products
- Glass
- Gypsum board
- Inert pigments
- Ivory
- Meats
- Metal parts.

You can see that buildings and structures of an accessory character and miscellaneous structures not classified in any specific occupancy must be constructed, equipped, and maintained to conform to the requirements of this code with the fire and life hazard to their occupancy.

GROUP U

There is one more group that is called Group U, a miscellaneous group that can include the following:

- Agricultural buildings
- Aircraft hangars
- Barns
- Carports
- Fences more than 6 feet high
- Grain silos, accessory to a residential occupancy
- Greenhouses
- Livestock shelters
- Private garages
- Retaining walls
- Sheds
- Stables
- Tanks
- Towers.

This brings us to the close of Chapter 1 and the occupancy classification of all buildings and structures. As you can see there are many exceptions that you must be aware of when constructing your building or structure. Read the code book, follow and implement the code, and if you ever have any questions, don't hesitate to contact your building official. They are there to help you.

Fast Fact

Take note that adult and child care facilities that are within a single-family home are permitted to comply with the International Residential Code.

CHAPTER

2

Special Detailed Requirements Based on Use and Occupancy

In Chapter 2 we will explore the provisions and apply them to the special uses of occupancy along with the occupancy and construction requirements of this code.

This section will apply to buildings or structures that are defined as covered mall buildings that do not exceed three floors at any point or not more than three stories above grade plane. Covered mall buildings must meet applicable provisions of this code except as specifically required. There are two exceptions. The first one states that foyers and lobbies of Groups B, R-1, and R-2 are not required to comply with this section. The second says that buildings do not need to comply with the section when they totally comply with other provisions of this code. Please note the word "totally," if your building does not comply with the fullest extent required then you will not be meeting the code standards.

For this chapter there are a few words and their definitions that I'd like to explore. They are used in this chapter and throughout this code book. They

!Definitionalert

Atrium: An opening connecting two or more stories, other than enclosed stairways, elevators, hoistways, escalators, plumbing, electrical, air-conditioning, or other equipment, which is closed at the top and not defined as a mall. Stories, as used in this definition, do not include balconies within assembly groups or mezzanines that comply with Chapter 5 of the code.

will not be written in text such as this, but you will find them as "Definition Alerts" throughout the book. Pay attention to them as they will come in handy and will be words that you need to be familiar with.

COVERED MALL BUILDINGS

The occupant load factor (OLF) is used in a formula to determine the means of exits of covered mall buildings and the number of occupants the exits is to be provided for. The means of exits are based on gross leasable area (GLA) of the covered mall building, but this excludes anchor buildings. The occupant load factor is determined by the following equation:

$$\text{OLF} = (0.00007)\,(\text{GLA}) + 25$$

The range for the OLF is not required to be less than 30 and must not exceed 50.

!Definitionalert

Covered mall building: A single building enclosing a number of tenants and occupants such as retail stores, drinking and dining establishments, entertainment and amusement facilities. This includes passenger transportation terminals, offices, and other similar uses where two or more tenants have a main entrance into one or more malls.

!Definitionalert

Anchor building: An exterior perimeter building of a group other than H having direct access to a covered mall building, but having the required exits independent of the mall.

Note the definition for covered mall building; I must tell you that for the purpose of this chapter, anchor buildings are not to be considered as part of the covered mall building.

For the purpose of determining the occupant load of a food court and exits, the food court occupant load needs to be added to the occupant load of the covered mall building as the calculation above.

If the travel used by any persons other than employees is more than 75 feet and your occupancy load is 50 people or more then you may not provide less than two exits. Assembly occupancies with an occupant load of 500 or more must be located in the covered mall building so that their entrance is immediately adjacent to a main entrance of the mall and must not have less than one-half of their required exits opening directly to the outside of the covered mall area. I said that the exits for the covered mall do not include anchor buildings.

Trade Tip

Anchor building exit requirements are different from the covered mall. The exits for anchor buildings must be provided independently form the malls means of exits. The occupant load of anchor buildings opening into the mall are not included in the determining means of exits for the mall either. Also the exit path of any mall cannot exit through anchor buildings. The mall is considered to be a dead-end mall when it ends at an anchor building and there are no other means of exit.

Did You Know?

The distance of exits for each individual tenant space in a covered mall building cannot exceed 200 feet. This travel is from any point to an exit or entrance and the maximum distance of travel from any point within a mall is also 200 feet.

Whether from an individual store or anywhere in the mall, there must be an exit no more than 200 feet away. Where more than one exit is required, they must be arranged so that is the travel in either direction from any point in the mall to separate exits. The passageway of an exit or corridor must be 66 inches. The exception to this is dead ends that exceed a length equal to twice the width of the mall measuring from the narrowest location within the dead end portion of the mall. Exit passageways providing a secondary means of exit from a tenant space must be protected by one-hour fire doors that are self- or automatic closing by smoke detection. Mechanical rooms, electrical room, building service areas, and service elevators are permitted to open directly into exit passageways, provided that the exit passageway is separated from these rooms with not less than one-hour fire-resistance-rated fire barriers and one-hour opening protective devices.

Attached garages used for the storage of passenger vehicles which have a capacity of nine or more passengers and open parking garages are considered separate buildings as they are separated from the covered mall building by a fire barrier having a fire-resistance rating or at least two hours. One exception applies: Where an open parking garage or enclosed parking

!Definitionalert

Gross leasable area: The total floor area designed for tenant occupancy and exclusive use. The area of tenant occupancy is measured from the centerlines of joint partitions to the outside of the tenant walls. All tenant areas, including areas used for storage, must be included in calculating gross leasable areas.

!Definitionalert

Food court: A public seating area located in the mall that serves food adjacent to tenant spaces.

garage is separated from the covered mall building or anchor building more than 10 feet away, exceptions may exist.

FIRE-RESISTANCE SEPARATIONS

Fire-resistance separations not only apply to parking garages and malls, but also to the tenants. Each tenant space must be separated from other tenant spaces by a fire partition that complies with the fire-resistance code section of this book. A tenant separation wall is not required between any tenant space and the mall. Anchor buildings need to be separated from the covered mall building unless the anchor building is not more than three stories above grade plane and has an occupancy classification the same that is permitted for tenants of the covered mall building that must be separated by two-hour fire-resistant fire barriers.

A covered mall building must be equipped throughout with a standpipe system as required by code. The code requires more than just fire-resistant walls and sprinkler systems for different areas of a covered mall building.

Fast Fact

According to code the minimum width of the mall must be 20 feet (6096 mm). The mall width must be sufficient enough to accommodate the occupant load. There needs to be a minimum of 10 feet (3048 mm) of clear exit width to a height of 8 feet (2438 mm) between any projection of a tenant space bordering the mall and nearest kiosk, vending machine, bench, food court, or other obstruction to the exit travel.

Did You Know?

Automatic sprinkler systems are required for covered mall buildings and buildings that are connected to them.

For example, kiosks or other similar structures (both temporary and permanent) cannot be located within the mall unless constructed of fire-retardant wood that complies with the code, foam plastics that have a maximum heat release rate not greater than 100kW (105 Btu/h) when tested in accordance with the exhibit booth protocol in UL 1975 or aluminum composite material (ACM) having a flame spread index of not more than 25 and a smoke-developed index of not more than 450 when tested as an assembly in the maximum thickness intended for use in accordance with ASTM E 84. All kiosks or similar structures located within the mall must also be provided with approved fire suppression and detection devices. Spacing within kiosks or similar structures must be at least 20 feet (6096 mm) and they must have a maximum area of 300 square feet. There are many structures that require special detailed requirements besides mall buildings or garages.

CHILDREN'S PLAYGROUNDS

Structures that are intended as children's playgrounds that exceed 10 feet in height and 150 square feet must comply with the following:

- Fire-retardant-treated wood
- Light-transmitting plastics complying with this code regarding plastics

Trade Tip

Except for the separation between Group R-1 sleeping units and a mall, openings between anchor buildings of Type IA, IB, IIA, and IIB construction and a mall do not need special fire protection.

CODE UPDATE

The minimum floor and roof opening width above grade shall be 20 feet in open malls.

- Foam plastics (including the pipe foam used in soft-contained play equipment structures) having a maximum heat-release rate not greater than 100kW when tested in accordance with UL 1975
- Aluminum composite material (ACM) meeting the requirements of Class A interior finish in accordance with Chapter 8 when tested as an assembly in the maximum thickness
- Textiles and films complying with the flame performance criteria in NFPA 701
- Plastic materials used to construct rigid components of soft-contained play equipment structures (such as tubes, windows, panels, and decks) meeting the UL 94 V-2 classification when tested in accordance with UL 94
- Foam plastics must be covered by a fabric, coating, or a film that meets the flame performance criteria of NFPA 701
- The interior floor under the children's playground structure must be a Class I interior floor finish classification when tested in accordance with NFPA 253.

Any playground structure for the use of children located within a mall must be provided with the same level of approved fire suppression and detection devices required for kiosks and similar structures. Children's playground structures are also required, just as kiosks are, to be separated from other structures. Playground structures are required to have a minimum horizontal separation of 20 feet from any other structures within the mall. They also have an area limit and cannot exceed 300 square feet unless a special investigation has demonstrated adequate fire safety.

SECURITY

One of the most important safety features of covered malls is security grilles and doors. Horizontal sliding or vertical security grilles or doors that are part of a required means of exit have to conform to rules as well. These rules are:

- They must remain in the full open position during the time periods that the mall is occupied by the general public.
- Doors or grilles cannot be closed when there are 10 or more persons occupying a space that has only one exit or when there are 50 or more persons occupying a space that has more than one exit.
- The doors or grilles must be able to be open from the inside of the building without the use of any special knowledge or special efforts where the space is occupied.
- Where two or more exits are required, not more than one-half of the exits are permitted to include a horizontal sliding or vertical rolling grille or door.

It is mandatory that covered mall buildings that exceed 50,000 square feet be provided with standby power systems that are capable of operating the emergency communication system. These emergency voice/alarm communication systems must be accessible to the fire department.

PLASTIC SIGNS

Plastic signs that are applied to the storefront of any tenant space facing the mall must adhere to the following rules in accordance with this code.

CODE UPDATE

Exits in malls are subject to the same fire resistance requirements as interior wall and ceiling finishes.

- Plastic signs must not exceed 20 percent of the wall area facing the mall.
- Plastic signs must not exceed a height of 36 inches (914 mm), except if the sign is vertical, the height cannot be higher than 96 inches (2438 mm), and the width cannot be wider than 36 inches (914 mm).
- Plastic signs must be located a minimum distance of at least 18 inches (457 mm) from adjacent tenants.
- Plastics other than foam used in signs must be light-transmitting plastics and shall have a self-ignition temperature of 650°F (343°C) or greater when tested in accordance with ASTM D 1929, and a flame spread index not greater than 75 and smoke-developed index not greater than 450 when tested in the manner intended for use in accordance with ASTM E 84.
- Edges and backs of plastic signs in the mall must be fully encased in metal.
- Foam plastics used in signs must have a maximum heat-release rate of 150 kilowatts when tested in accordance with UL 1975 and the foam plastics must have the physical characteristics specified in this section. Foam plastics used in signs installed in accordance with this section are not required to comply with the flame spread and smoke-developed indexes specified in plastics.
- The minimum density of foam plastics used in signs must not be less than 20 pounds per cubic foot (pcf) and the thickness must not be greater than ½ inch (12.7 mm).

HIGH-RISE BUILDINGS

The provisions of this section apply to buildings with an occupied floor located more than 75 feet above the lowest level of fire department vehicle access. There are five exceptions to these provisions which include the following:

- Airport traffic control towers
- Open parking spaces

CODE UPDATE

Malls that contain atriums must be equipped with smoke-control systems.

- Buildings with occupancy in Group A-5
- Low-hazard special industrial occupancies
- Buildings with occupancy in Group H-1, H-2, or H-3.

Buildings and structures must be equipped throughout with an automatic sprinkler system and a secondary water supply where required. We know from earlier discussion that open parking garages are an exception to this provision. A reduction in fire-resistant rating is allowed in buildings that have sprinkler control valves equipped with supervisory initiating devices and water flow initiating devices for each floor. For buildings not greater than 420 feet in height, Type IA construction is allowed to be reduced to Type IB with the exception of columns supporting the floor which may not be allowed to be reduced.

For buildings not greater than 420 feet in height, the required fire-resistance rating of the fire barriers enclosing vertical shafts, other than exit enclosures and elevator hoistway enclosures, must be reduced to one hour where automatic sprinklers are installed within the shafts at the top and at alternate floor levels. If standby power is a generator and is set inside a building, the generator must be located in a separate room that is enclosed with two-hour fire barriers. Any systems you use with a manual ignition and transfer features have to be provided at the fire command center.

Standby power loads can be any of the following: power and lighting for the fire command center, electrically powered fire pumps and ventilation, and automatic fire detection equipment for smokeproof enclosures. All elevators must have standby power. The following is a list of what is classified as an emergency power load:

- Exit signs and means of exits that are illuminated
- Elevator car lighting

- Emergency voice/alarm communication systems
- Automatic fire detection systems
- Fire alarm systems.

UNDERGROUND BUILDINGS

In this section I will touch upon underground building spaces and the codes that apply to them. The building spaces that I refer to have a floor level used for human occupancy more than 30 feet below the lowest level of exits. There are a few exceptions to this. They are as follows:

- One- and two-family dwellings, having a compliant sprinkler system
- Parking garages with compliant automatic sprinkler systems
- Fixed guide way transit systems
- Grandstands, bleachers, stadiums, arenas, and similar facilities
- When the lowest story of a building is the only one that would qualify the building as an underground building. That building must not have an area that exceeds 1500 square feet and have an occupant load that is less than 10.

The underground portion of a building must be of Type I construction and the highest level of exit must be equipped with an automatic sprinkler system with water-flow valve switches and control valves that are supervised.

Underground buildings that have a floor level or more than 60 feet (18,288 mm) below the level of exit must be divided into two compartments. These compartments must be of an approximately equal size and extend through the highest level of exit discharge that serve the underground portion of the building and the levels below. As you may have guessed, there is an exception to this. If the area does not exceed 1500 square feet and has an occupant load of less than 10, the lowest story does not need to be compartmented. Each compartment needs to be separated by a smoke barrier wall and penetrations between the compartments will be limited to plumbing and electrical piping and conduit that are fire stopped.

Doorways must be protected by fire door assemblies that close automatically when smoke alarms go off. When provided, each compartment must

CODE UPDATE

Atriums must be equipped with fire alarm systems.

have an air supply and an exhaust system that is independent of other compartments. Doors must be installed in accordance with NFPA 105. If elevators are provided, each compartment must have access to an elevator. If an elevator serves more than one compartment, an elevator lobby must be provided and be separated by a smoke barrier. Make sure that when installing elevators the doors are gasketed and have a drop sill. They must also have automatic closing by smoke detection.

When compartmentation is required, each compartment needs to have an independent smoke control system. It must be automatically activated and capable of manual operation. Underground buildings can require both fire alarm systems and a public address system. Each floor of an underground building must have a minimum of two exits; if compartmentation is required then the compartments must have a minimum of one exit and have an exit access doorway into an adjoining compartment.

Staircases in underground buildings that serve floor levels of more than 30 feet below their level of exit must comply with the requirements for a smoke proof enclosure. It is important that your underground building have standby power and emergency power for blackouts, storms, or other situations which cause power outages. Having these systems in place can save lives and make exiting the building easier. Listed below are loads which classify as standby power loads:

- Smoke control system
- Ventilation and automatic fire detection equipment for smokeproof enclosures
- Fire pumps.

The standby power system must start up within 60 seconds after the failure of normal power supply. As I have listed loads for standby power, I will do so for emergency power loads as well.

- Emergency voice/alarm communication systems
- Fire alarm systems
- Automatic fire detection systems
- Elevator car lighting
- Means of exits and exit sign illumination.

You must have a standpipe system equipped throughout the underground building system in accordance with the code.

GROUP I-2 MEDICAL FACILITIES

Corridors in occupancies in Group I-2 must be continuous to the exits and separated from other areas in accordance with corridor walls except spaces conforming to the following paragraphs. Spaces of unlimited areas such as waiting areas and similar spaces constructed for corridors must be permitted to be open to a corridor, but only when the following requirements are met:

- The spaces are not occupied for patient sleeping units, treatment rooms, hazardous, or incidental use in mixed uses and occupancy.
- The open space is protected by an automatic fire detection system installed in accordance with the code.
- The corridors onto which the spaces open, in the same smoke compartment, need to be protected by an automatic fire detection system installed. Smoke compartment located in the spaces are to be equipped throughout with quick-response sprinklers.
- The space is arranged so that it does not obstruct the exit.

Trade Tip

Spaces for doctors and nurses used for charting, communications, and other clerical areas have to be open to the corridor when these spaces are constructed as required for corridors.

Accordance for mental health treatment areas require that mental health patients who are not capable of self-care and are housed in spaces for group meetings or multipurpose therapeutic spaces other than incidental use areas, and under continuous supervision by facility staff, must be permitted to be open to the corridor, where the following criteria must be met:

- Each area must not exceed 1500 square feet.
- The area is to be located so that supervision by the facility staff is permitted.
- The area is to be arranged so that access to exits is not obstructed in any way.
- The area is to be equipped with an automatic fire detection system.
- Not more than one such space is permitted in any one smoke compartment.
- The walls and ceilings of the space are to be constructed as required for corridors.

In these types of facilities there is almost always a gift shop where visitors and patients may go to purchase small items such as gum, flowers, daily newspapers or magazines, if permitted. These gift shops must be less than 500 square feet in area and must be permitted to be open to the corridor provided the gift shop and storage areas are fully equipped with sprinklers and the storage areas are protected in accordance with this code.

Corridor walls must be constructed as smoke partitions. Doors, other than those in a wall required to be rated or for the enclosure of a vertical opening or an exit, shall not have a required fire protection rating and are not required to be equipped with self-closing or automatic-closing devices. These automatic-closing devices must provide an effective barrier to limit

CODE UPDATE

An underground building is one that extends more than 30 feet below the finished floor of the lowest level of exit discharge.

smoke and cannot be equipped with positive latching. Roller latches are not permitted.

When locks are in place and restrict access to the patient room from the corridor and can only be used by the staff from the corridor side, they cannot obstruct or restrict any means of exits from the patient room except for patient rooms in mental health facilities.

Smoke barriers in health facilities must be provided to subdivide every story used by patients for sleeping or treatment. Smoke barriers must also divide other stories with an occupant load of 50 or more persons into at least two smoke compartments. These stories have to be divided in smoke compartments with an area of not more than 22,500 square feet and the travel distance from any point in a smoke compartment to a smoke barrier door must not exceed 200 feet. Should the need arise, a refuge area must be provided. At least 30 net square feet per patient must be provided within the aggregate area of corridors, patient rooms, treatments rooms, lounge or dining areas and other low-hazard areas on each side of each smoke barrier. On the floors that do not have patients confined to a bed, at least 6 net square feet per occupant must be provided on each side of each smoke barrier for the total number of occupants in adjoining smoke compartments.

All smoke compartments have to provide an independent means of exit so that staff, residents, patients, and family do not have to return through the smoke compartment to get to safety. Automatic sprinkler systems must be installed in smoke compartments that contain patient sleeping areas. The smoke compartments must also be equipped with an approved quick-response or residential sprinklers.

Automatic fire detection for corridors in nursing homes, both intermediate and skilled nursing facilities, detoxification facilities, and spaces permitted to be open to the corridors which I mentioned as an independent exit must also be equipped with an automatic fire detection system. Hospitals also must have smoke detection installed as required. There are two exceptions:

- Corridor smoke detection is not required where patient sleeping units are provided with smoke detectors that comply with UL 268. These detectors must provide a visual display on the corridor side of each patient sleeping unit and an audible and visual alarm at the nursing station that monitors each unit.

!Definitionalert

Sally port: A small controlled space with two doors. One must enter the space and close the first door before opening the second door to proceed.

- Corridor smoke detection is not required where patient sleeping unit doors are equipped with automatic door-closing devices with integral smoke detectors on the inside of the units. These must be installed in accordance with their listing, provided that the detectors perform the required alerting function.

Some facilities have secured yards installed for the safety of the patients and staff. Grounds are permitted to be fenced and gates are permitted to be equipped with locks provided that safe dispersal areas having 30 net square feet for bed and litter patients and 6 net square feet for patients who can walk and other occupants are located between the building and the fence. These safe dispersal areas cannot be located less than 50 feet from the building that they are used for.

I-3 DIFFERENT OCCUPANCIES

Not all facilities that house patients for various reasons are necessarily medical. There are buildings and structures that occupy patients for criminal activities. Where security operations need the locking of required means of exits, provisions must be made for the release of occupants at all times. Exits from detention and correctional occupancies that pass

Did You Know?

Facilities that have spiral stairs are permitted for access to and between staff locations.

through other areas of use must, as a minimum, conform to requirements for detention and correctional occupancies. It is possible to exit through a horizontal exit into other adjoining occupancy exit provisions that do comply with requirements used in the appropriate occupancy, as long as the occupancy is not a high-hazard use. The means of exit must comply with the code book except as modified or as provided for in this section. Doors to resident sleeping areas need to have clear width of no less than 28 inches. If the door used for an exit is a sliding door, the force to slide the door fully open cannot exceed 50 pounds with an upright force against the door of 50 pounds.

Facilities may have exit doors that lead into a fenced or walled courtyard. These courtyards must be of a size that accommodates all occupants with a minimum of 50 feet from the building with a net area of 15 square feet per person. During an emergency at a correctional facility a sally port may be used as a means of exits when there are provisions for continuous and unobstructed passage through the sally port during an emergency exit condition. Of all the required exit enclosures in each building, one of them may be permitted to have glazing installed in each door and inside walls at each landing level providing access to the enclosure, only if the following conditions are met:

- The exit enclosure cannot serve more than four floor levels.
- Exit doors cannot be less than ¾-hour fire door assemblies.
- The total area of glazing at each floor level cannot exceed 5000 square inches and individual panels of glazing cannot exceed 1296 square inches.
- The glazing will be protected on both sides by an automatic fire sprinkler system. The sprinkler system will be designed to completely soak the entire surface of any glazing affected by fire.
- The glazing must be in a gasketed frame and installed in such a manner that the framing system will deflect without breaking the glass before the sprinkler system turns on.
- Obstructions, such as curtain rods, drapery rods, curtains, drapes, or similar materials will not be installed between the automatic sprinklers and the glazing.

Exit doors are allowed to be locked in accordance with the appropriate use. Doors from an area of protection to the outside are permitted to be locked with a key. The keys to unlock the outside doors must be available at all times and the locks must be operable from both sides of the door. If the locks on the doors to exits are of a remote release, they must be provided with a reliable means of operation. There is a great risk that one is taking in their own hands if the remote release does not work due to poor maintenance or by any other means. Human lives can be lost if these types of locks are not checked periodically for bugs in the system. If at any time you find that these types of locks are not working, report it to your supervisor immediately so that it may be fixed in a timely manner. Think about this, would you prefer the building inspector or an employee to find this potentially dangerous matter?

In occupancy Conditions III and IV, the arrangement, accessibility, and security of the release mechanism(s) required for exit must have the minimum available staff at any time, the lock mechanisms are capable of being released within two minutes. Provisions for remote locking and unlocking of occupied rooms in Occupancy Condition IV are not required provided that not more than 10 locks are necessary to be unlocked in order to move occupants from one smoke compartment to a safe area within three minutes.

Opening of necessary locks must be accomplished with not more than two separate keys. Power-operated doors and locks are very different from remote release and as the owner, manager, or supervisor of Group I-3 occupancy, it is important that you know what the difference is and understand the use of the locking systems. Let us take a look at them now.

Power-Operated Sliding Doors

Power-operated sliding doors or power-operated locks for swinging doors must be used by a manual release mechanism at the door, and either emergency power or a remote operating release must be provided. But, emergency power is not required in facilities with 10 locks or less when complying with the exception of the codes for remote release.

Remote release, mechanically operated sliding doors or remote release, mechanically operated locks must be provided with a mechanically operated release mechanism at each door, or must be provided with a redun-

CODE UPDATE

Stairways in underground buildings must be equipped with a smoke-proof enclosure.

dant remote release control. Doors that are unlocked remotely under emergency conditions must not automatically relock when closed unless specific action is taken at the remote location to enable the door to relock.

Vertical Openings

Vertical openings must be enclosed with these exceptions:

- An opening in the floor, between floor levels of residential housing areas, is allowed without enclosure protection between the levels provided that the entire, normally occupied areas that are interconnected are open and unobstructed to allow supervisory personnel visual access to the areas.
- Exit space is plenty enough to provide immediate exits for all occupants from all connected levels and areas.
- The height difference between the highest and lowest finished floor levels cannot exceed 23 feet (7010 m).
- Each level will be considered separately.
- Occupancies in Group I-3 also must have smoke barriers that divide every floor occupied by residents for sleeping, or any floor having an occupant load of 50 or more persons. The division must be made into at least two smoke compartments.

There is an exception to the guidelines above. Spaces having a direct exit to one of the following, provided that the locking arrangements of the doors involved comply with the requirements for doors at the smoke barrier for the condition involved to be:

- A public way

- A building separated from the resident housing area by a two-hour fire-resistance-rated assembly or 50 feet (15,240 mm) of open space.
- A secured yard or court having a holding space 50 feet from the housing area that proved 6 square feet or more of safety area per occupant. This includes residents, staff, and visitors.

Smoke compartments require that the maximum number of residents in any smoke compartment will be 200. The travel distance to a door in a smoke barrier from any room door required as exit access will not be longer than 150 feet. The distance to a door in a smoke barrier from any point in that room cannot be longer than 200 feet. In the event of an emergency, such as fire, not only is it important to have a safe place for people to flee to, but it is a code law. Facilities housing detention and correctional occupancies require at least 6 net square feet per occupant. This space must be readily available wherever the occupants are moved across the smoke barrier in a fire emergency.

An exit must be provided from each compartment that is created by smoke barriers without having occupants escape through the smoke compartment by the original exits. Resident housing areas and sleeping areas and any nearby day rooms, group activity space, or other common spaces where residents live must be separated from other spaces in accordance with the following:

- Occupancy Conditions III and IV: Each sleeping area in these conditions must be separated from nearby common spaces by a smoke-tight partition where the distance from the sleeping area through the common area to the corridor is more than 50 feet.
- Occupancy Condition V: Each sleeping area in occupancy condition V will be separated from nearby sleeping areas, corridors, and common spaces by a smoke-tight partition.
- Openings in rooms: The combined area must include all openings, such as door undercuts, food passes, and grilles. Openings cannot be more than 36 inches above the floor and the openings must be closeable from the room side.

!Definitionalert

Fly gallery: A raised floor area above a stage from which the movement of scenery and operation of other stage effects are controlled.

- Smoke-tight doors in openings in partitions required to be smoke tight as above must be substantial doors, with construction that will resist the passage of smoke.
- Latches and door closures are not required on cell doors.

For the purposes of this section, a windowless building or part of a building is one with windows that do not open, break easily, or without windows. Windowless buildings must be provided with an engineered smoke control system to provide ventilation. This can be either mechanical or natural, but must be in accordance with the code for each windowless compartment.

STAGES AND PLATFORMS

This section includes provisions to all parts of buildings and structures that contain stages or platforms and similar additions as defined.

A stage is defined as a space within a building used for entertainment or presentations, which includes overhead hanging curtains, drops, scenery or stage effects other than lighting, and sound. Stages must be constructed of materials as required for floors for the type of construction of the building in which such stages are located. The following list contains exceptions to this.

Fast Fact

Floors of fly galleries and catwalks must be constructed of any approved material.

!Definitionalert

Gridiron: The structural framing over stage supporting equipment for hanging or flying scenery and other stage effects.

- Stages of Type IIB or IV construction with a normal 2-inch (51 mm) wood deck, provided that the stage is separated from other areas in accordance with this section.
- In buildings of Type IIA, IIIA, and VA construction, a fire-resistance-rated floor is not required, provided the space below the stage is equipped with an automatic fire-extinguishing system.
- In all types of construction, the finished floor must be constructed of wood or approved noncombustible materials.
- Openings through stage floors must be equipped with tight-fitting, solid wood trap doors with approved safety locks.
- Stage height and area must be measured to include the entire performance area and adjacent backstage and support areas not separated from the performance area by fire-resistance-rated construction.
- Stage height must be measured form the lowest point on the stage floor to the highest point of the roof or floor deck above the stage.

Galleries, gridirons, catwalks, and pinrails must use beams that are designed for this purpose only, the beams must be constructed of approved materials consistent with the requirements for the type of construction of the buildings, and a fire-resistance rating is not required. The areas will not be considered to be floors, stories, mezzanines, or levels when applying this code; an exception to remember is this: Floors of fly galleries and catwalks must be constructed of any approved material.

Stage Doors

Exterior stage doors must be protected with fire door assemblies. Exterior openings that are located on the stage for means of exit or loading and unloading purposes, and are likely to be open during occupancy of the theater, must be constructed with vestibules to prevent air drafts into the theater. In the event that the stage height is greater than 50 feet, all portions of the stage must be completely separated from the seating area by a wall with no less than a two-hour fire-resistance rating extending continuously from the foundation to the roof.

Where the separated wall is required to have a fire-resistance rating, the stage opening must have a fire curtain of approved material or an approved water curtain that complies with the code. The fire curtain must be designed and installed to intercept hot gases, flames, and smoke and to prevent a glow from a severe fire on the stage from showing on the auditorium side for a period of 20 minutes. The closing of the fire curtain from the full open position must be accomplished in less than 30 seconds, with the last 8 feet of travel requiring five or more seconds for full closure. The curtain must be activated by rate-of-rise heat detection installed in accordance with the code, operating at a rate of temperature rise of 15 to 20°F per minute, and by a secondary manual control.

A sample curtain with a minimum of two vertical seams must be subjected to the standard fire test specified in ASTM E 119 for a period of 30 minutes. The curtain will overlap the furnace edges by an amount that is appropriated to seal the top and sides. There will be a bottom pocket on the curtain that contains a minimum of 4 pounds per linear foot of batten. The exposed surface of the curtain mustn't glow, and flame or smoke must not penetrate the curtain during the test period. Unexposed surface temperature and hose stream test requirements are not applicable to the wall fire safety curtain test. The fabric of the curtains that are used in such cases must have a smoke-developed rating of 25 or less when tested in accordance with ASTM E 84. The completed curtain that is in place between the stage and the wall will be subjected to operating tests prior to the issuance of a certificate of occupancy.

CODE UPDATE

Egress doors are permitted to be locked in accordance with the applicable use condition.

COMBUSTIBLE MATERIALS

Most plays and theater production use some kind of scenery or props to enhance performances. Combustible materials used in sets and scenery must meet the fire movement performance criteria of NFPA 701 and the International Fire Code. Foam plastics and materials containing foam plastics must comply with the chapter regarding plastics and the International Fire Code.

Emergency ventilation must be provided for stages larger than 1000 square feet in floor area, or with a stage height greater than 50 feet. This ventilation must comply with one or both of the requirements of the code.

Roof Vents

Two or more roof vents constructed to open automatically by approved heat-activated devices and with an aggregate clear opening area of not less than 5 percent of the area of the stage can be located near the center and above the highest part of the stage area. Supplemental means must be provided for manual operation of the ventilator. Curbs must be provided as required for skylights and vents must be labeled.

Smoke control must be provided to maintain the smoke layer crossing point not less than 6 feet above the highest level of the assembly seating or above the top of the space where the stage is separated from the seating area or where a wall is provided in this space.

Did You Know?

The code is much different for temporary platforms and states that platforms installed for a period of not more than 30 days are permitted to be constructed of any materials permitted by the code. But, the space between the floor and the temporary platform can only be used for plumbing and electrical wiring to platform equipment. Be sure that you know the difference in codes for permanent and temporary platforms.

Did You Know?

Stage exits must have at least one approved means of exit from each side of the stage and from each side of space under the stage. At least one of these exits must be provided from each fly gallery and from the gridiron. Steel ladders, alternating tread stairways or spiral stairways are means of exits that are permitted to be used from the gridiron to a scuttle in the stage roof.

Platform Construction

Platform construction for permanent platforms must be constructed of materials as required for the type of construction of the building where your platform is going to be located. Permanent platforms are allowed to be constructed of fire-retardant-treated wood for Type I, II, and IV construction where the platforms are no more than 30 inches above the main floor, and are not more than one-third of the room floor area and not more than 3000 square feet in area. If the space beneath the permanent platform is used for storage or any other purpose other than equipment, wiring, or plumbing, the floor construction cannot be less than one-hour fire-resistance-rated construction. If the space beneath the permanent platform is used only for equipment, wiring, or plumbing, the underside of the permanent platform does not need to be protected.

Dressing Rooms

All dressing rooms and appurtenant rooms must comply with the following section. If the stage height is greater than 50 feet, the stage needs to be separated from dressing rooms, scene docks, property rooms, workshops, storerooms, and compartments that are an extension to the stage and other parts of the building by a fire barrier with no less than a two-hour fire-resistance rating with approved opening protections. If the stage is 50 feet or less, the required stage separation must be a fire barrier with no less than a one-hour fire-resistance rating with approved protective openings. You must make all dressing rooms, scene decks, property rooms, workshops, storerooms, and compartments that are extended to the stage separated from each other as well. They must have fire barriers with not less than a one-hour fire-resistance rating with approved protected openings.

Automatic Sprinkler Systems

You have learned from the previous paragraphs that all buildings or structures require an automatic sprinkler system. The codes for each building are not quite the same, so it is very important that you don't apply one section to another. The automatic sprinkler code for stages and platforms is the following: Stages must be equipped with an automatic fire-extinguishing system. Sprinklers must be installed under the roof and gridiron and under all catwalks and galleries over the stage. Sprinklers must be installed in dressing rooms, performer lounges, shops, and storerooms accessory to the stages. The exceptions to this are that sprinklers are not required under stage areas less than 4 feet in clear height that are used exclusively for storage of table and chairs, provided that the concealed space is separated from the adjacent spaces by no less than ⅝-inch Type X gypsum board. Sprinklers are also not required for stages that are 1000 square feet or less in area and 50 feet or less in height where curtains, scenery, or other combustible hangings are not vertically retractable. Combustible hangings must be limited to a single main curtain, border, legs, and a single backdrop. Sprinklers are also not required within portable orchestra enclosures on stages.

AIRCRAFT-RELATED OCCUPANCIES

This section and the provisions of this section will apply to airport traffic control towers not exceeding 1500 square feet per floor occupied for the following uses only:

!Definitionalert

Heliport: An area of land or water or a structural surface that is used, or intended for use, for the landing and taking off of helicopters, and any accessory areas that are used, or intended for use, for heliport buildings and other heliport facilities.

- Airport traffic control cab
- Electrical and mechanical equipment rooms
- Airport terminal radar and electronic rooms
- Office spaces incidental to tower operation
- Lounges for employees, including sanitary facilities.

A minimum of one exit stairway must be permitted for airport traffic control towers of any height provided that the occupant load per floor does not exceed 15. The stairway must be separated from elevators by a minimum distance of one-half of the diagonal of the area served measured in a straight line. The exit stairways and elevator hoistways are permitted to be located in the same shaft enclosure, provided they are separated from each other by a four-hour fire barrier having no openings. Such stairways must be pressurized to a minimum of 0.15 inch of water columns and a maximum of 0.35 inch of water column in the shaft relative to the building with stairway doors closed.

Stairways need not extend to the roof and the provisions of high-rise buildings do not apply. If required stairways are pressurized, smokeproof enclosures do not apply. Airport traffic control towers must be provided with an automatic fire detection system installed. A standby power system that conforms to the electrical section of this book must be provided in airport traffic control towers more than 65 feet in height.

Power must be provided to pressurization equipment, mechanical equipment and lighting. This must also include elevator operating equipment and fire alarms and smoke detections systems. Airport traffic control towers do not need to be accessible.

HANGERS

Aircraft hangars must be in accordance with the following:

- Exterior walls located less than 30 feet from property lines, lot lines, or a public way must have a fire-resistance rating of no less than two hours.
- Where hangars have basements, the floor over the basement must be of Type IA construction and must be made tight against seepage or water, oil, or vapors.

- There will be no opening or communication between a basement and a hangar.
- Access to a basement will be from the outside only.
- Floors must be graded and drained to prevent water or fuel from staying on the floor.
- Floor drains must discharge through an oil separator to the sewer or to an outside vented sump. An exception to this is aircraft hangars with individual lease spaces not exceeding 2000 square feet each in which servicing, repairing, or washing is not conducted and fuel is not dispensed. These must have floors that are graded toward the door, but do not require a separator.
- All heating equipment must be placed in another room that is separated by two-hour fire-resistance-rated construction.
- Entrance must be made from the outside or be means of a vestibule providing a two-doorway separation. There are, however, some exceptions.

Unit heaters and vented infrared radiant heating equipment suspended at least 10 feet above the upper surface of wings or engine enclosures of the highest aircraft that are permitted to be housed in the hangar and at least 8 feet above the floor in shops, offices, and other sections of the hangar communicating with storage or service areas.

A single interior door must be allowed, provided the sources of ignition in the appliances are at least 18 inches above the floor.

!Definitionalert

Residential aircraft hangars: An accessory building less than 2000 square feet (186 m2) and 20 feet (6096 mm) in height, constructed on a one- or two-family residential property where aircraft are stored. Such use will be considered as a residential accessory use incidental to the dwelling.

Fast Fact

A hangar must provide two means of exits. One of the doors into the dwelling must be considered as meeting only one of the two means of exits.

The process of "doping," involving use of a volatile flammable solvent, or of painting, must be carried on in a separate detached building equipped with automatic fire-extinguishing equipment.

Fire suppression must be provided to aircraft hangars as required by NFPA 409, with the exception that Group II hangars as defined in NFPA 409 storing private aircraft without major maintenance or overhaul are exempt from foam suppression requirements.

A hangar cannot be attached to a dwelling unless separated by a fire barrier having a fire-resistance rating of no less than one hour. This separation must be continuous from the foundation to the underside of the roof and un-pierced except for doors leading to the dwelling unit. Doors into the dwelling unit must be equipped with self-closing devices with at least a 4-inch noncombustible raised sill. Openings from a hangar directly into a room used for sleeping purposes are not permitted.

Smoke alarms must be provided within the hangar in accordance with the code. All electrical, mechanical, and plumbing drain, waste, and vent or DWV systems installed within the hangar must be independent of the systems installed inside the home. Building sewer lines will be permitted to be connected outside the structures with the exception of smoke detector wiring and feed for electrical sub panels in the hangar.

Any aircraft painting operations where the use or storage of flammable liquids in excess of the maximum allowable quantities per control area have to be conducted in an aircraft paint hangar that complies with this section. Spray equipment and cleaning operations must be conducted in a liquid use, dispensing and mixing room.

Aircraft paint hangars must be classified as Group H-2. Aircraft paint hangars must comply with the applicable requirements of this code and the International Fire Code for such occupancy. The construction of such paint hangars will be of Type I or II construction.

All flammable liquids must be stored in a liquid storage room, and aircraft paint hangars must be provided with fire suppression as required by NFPA 409; all aircraft paint hangars will be provided with ventilation as required by the International Mechanical Code.

COMBUSTIBLE STORAGE

Combustible storage is always a potential concern. High-pile stock or rack storage in any occupancy group will comply with the International Fire Code. Attic, under-floor, and concealed spaces used for storage of combustible materials will be protected on the storage side as required for one-hour fire-resistance-rated construction. Openings will be protected by assemblies that are self-closing and are of non-combustible construction on solid wood core no less than 1¾ inches in thickness. The exceptions to this include areas that are protected by automatic sprinkler systems and Group R-3 and U occupancies.

HAZARDOUS MATERIALS

Buildings and structures with occupancy in Group H, that house hazardous materials, will also comply with the applicable provisions of this section and the International Fire Code.

The safe design of hazardous material occupancies is material dependent. Level 2 and 3 aerosol products will be stored and displayed in accordance with the International Fire Code for occupancy group requirements.

Separate floor plans must be submitted for buildings and structures with occupancy in Group H, identifying the locations of anticipated contents and processes to reflect the nature of each occupied portion of every building and structure. A report identifying hazardous materials including, but not limited to, materials that represent hazards that are classified in Group H to be stored or used, must be submitted and the methods that you plan on storing and protecting these hazards have to be indicated on the construction documents. The opinion and report has to be prepared by a qualified person, firm, or corporation by the building official and will be provided free of charge to the enforcing agency.

Control areas need to be kept separate from each other by fire barriers or horizontal assemblies, or both. The percentage of maximum allowable quantities of hazardous materials per control area permitted at each floor will be in accordance with code recommendations, as do the maximum number of control areas per building and the fire-resistance-rating requirements. There are two exceptions for the fire-resistance-rating requirements. These are: The floor construction of the control area and the construction supporting the floor of the control area are allowed to be one-hour fire-resistance rated in buildings of Type IIA, IIIA, and VA construction provided that the building is equipped throughout with an automatic sprinkler system and the building is three stories or less in height.

The aggregate quantity of nonflammable solid and nonflammable or noncombustible liquid hazardous materials permitted in Group M display and storage area, a Group S storage area, or an outdoor control area is permitted to exceed the maximum allowable quantities per control area specified in the code without classifying the building or use as a Group H occupancy, provided that the materials are displayed and stored in accordance with the International Fire Code and quantities do not exceed the maximum in this code.

Trade Tip

Any ducts that convey explosives or flammable vapors, fumes, or dusts need to extend directly to the outside of the building without entering other spaces. Exhaust ducts cannot and will not extend into or through ducts and plenums except ducts conveying vapor or fumes having flammable constituents less than 25 percent of their lower flammable limit (LFL) are permitted to pass through other spaces. Factories that have emissions generated at workstations must be confined to that area in which they are generated as specified in the International Fire Code and International Mechanical Code.

In Group M occupancy wholesale and retail sales uses, indoor storage of flammable and combustible liquids must not exceed the maximum allowable quantities per control area as indicated in this code.

The maximum quantity of aerosol products in Group M occupancy retail display areas and retail storage areas must be in accordance with the International Fire Code.

Ventilation of any rooms, areas, or spaces of Group H in which explosive, corrosive, combustible, flammable, or highly toxic dusts, mists, fumes, vapors, or gases are or may be emitted due to the processing, use, handling, or storage of materials must be mechanically ventilated as required of the International Fire and International Mechanical Codes.

Supply and exhaust openings must be located in accordance with the International Mechanical Code and exhaust air that is contaminated by highly toxic material must be treated in accordance with the International Fire Code. A manual shutoff control for ventilation equipment required by this section must be provided outside the room adjacent to the main access door to the room. The switch must be breakable and will be labeled: Ventilation System Emergency Shutoff.

If a building or structure has systems involving hazardous materials, the systems must be suitable for the intended applications. These systems are not interchangeable, they are designed to meet the individual designs and should not be changed around to suit your needs. It is your responsibility to obtain the proper systems for your needs. Controls must be designed to prevent materials from entering or leaving process or reaction systems at other than the intended time, rate, or path. Automatic controls, where provided, must be designed to be fail-safe.

Did You Know?

Monitor control equipment must be installed where required by the International Fire Code, and Group H occupancies must be provided with an automatic fire detection system.

Did You Know?

Protection from weather is a very important issue in regards to hazardous materials. Where weather protection is provided for sheltering outdoor hazardous material storage or use areas, such areas must be considered outdoor storage or use when the weather protection structure complies with the following section.

Explosion control must be provided in accordance with the International Fire Code where quantities of hazardous materials exceed the maximum allowable quantities or where a structure, room, or space is occupied for purposes involving explosion hazards as required by this chapter or the International Fire Code.

There must be a standby or emergency power source should an emergency or crisis happen. Where mechanical ventilation, treatment systems, temperature control, alarm, detection, or other electrically operated systems are required, such systems must be provided with an emergency or standby power system in accordance with this code or the ICC Electrical Code. The following are exceptions:

- Storage areas for Class I and II oxidizers
- Storage areas for Class III, IV, and V organic peroxides
- Storage, use, and handling areas for highly toxic or toxic materials as provided for in the International Fire Code
- Standby power for mechanical ventilation, treatment systems, and temperature control systems are not required where an approved fail-safe engineered system is installed.

Rooms, buildings, or areas occupied for the storage of solid and liquid hazardous materials must be provided with a means to control spillage and to contain or drain off the spillage and fire protection water discharged in the storage area where required in the International Fire Code, this is to include the methods of spill control and outdoor storage, dispensing, and use of hazardous materials.

Walls must not obstruct more than one side of the structure unless the obstructed area does not exceed 25 percent of the structure's perimeter. The distance from the structure to buildings, lot lines, pubic ways, or means of exits cannot be less than the distance required for an outside hazardous material storage or use area without weather protection. The overhead structure must be of approved noncombustible construction with a maximum area of 1500 square feet.

Storage

An approved manual emergency system must be provided in buildings, rooms, or areas used for storage of hazardous materials. Emergency alarm-initiating devices will be installed outside of each interior exit or exit access door of storage buildings, rooms, or areas. The activation of an alarm-initiating device must sound a local alarm to alert occupants of an emergency situation involving hazardous materials.

Hazardous Materials with Rankings of 3 or 4

Where hazardous materials having a hazard ranking of 3 or 4 in accordance with NFPA 704 are transported through corridors or exit enclosures, there must be an emergency telephone system, a local manual alarm station, or an approved alarm-initiating device at not more than 150-foot intervals and at each exit and exit access doorway throughout the transport route. The signal must be relayed to an approved central, proprietary, or remote station service or a location that is constantly attended or supervised; this must also initiate a local audible alarm.

The next section on hazardous materials is a lengthy one, but one that cannot be taken too seriously This next section is about hazardous materials and covers Groups H-1, H-2, H-3, H-4, and H-5.

OCCUPANCY GROUPS H-1, H-2, H-3, H-4, AND H-5

There are several groups pertaining to hazardous materials. The provisions of this section apply to the storage and use of hazardous materials in excess of the maximum allowable quantities per control area. Buildings and structures with occupancy in Group H will also comply with this section and the International Fire Code.

!Definitionalert

Continuous gas-detection system: A gas detection system where the analytical instrument is maintained in continuous operation and sampling is performed without interruption. Analysis is allowed to be performed on a cyclical basis at intervals not to exceed 30 minutes.

Group H occupancies must be located on property in accordance with other provisions of this chapter. In Groups H-2 and H-3, not less than 25 percent of the perimeter wall of the occupancy will be an external wall. The following are exceptions to this:

- Liquid use, dispensing, and mixing rooms having a floor area or not more than 500 square feet need not be located on the outer perimeter of the building where they are in accordance with the International Fire Code and NFPA 30.
- Liquid storage rooms having a floor area of not more than 1000 square feet need not be located on the outer perimeter where they are in accordance with the International Fire Code and NFPA 30.
- Spray paint booths that comply with the International Fire Code need not be located on the outer perimeter.

Group H occupancy minimum fire separation distance: regardless of any other provisions, buildings containing Group H occupancies must be set back to the minimum fire separation distance as described below. Distances must be measured from the walls enclosing the occupancy to lot lines, including those on public property. You may not use distances to assumed lot lines to establish the minimum fire separation for buildings on sites where explosives are manufactured or used when separation is provided in accordance with the quantity distance tables specified for explosive materials in the International Fire Code.

Group H-1 cannot be less than 75 feet and not less than required by the International Fire Code except for fireworks manufacturing buildings separated in accordance with NFPA 1124, buildings containing organic peroxides, unclassified detonable, and unstable reactive materials of class III and IV that are detonable and detonable pyrophoric materials, floors in storage areas for these materials must be of liquid-tight noncombustible construction. Group H-2 cannot be less than 30 feet where the area of the occupancy exceeds 1000 square feet and it is not required to be located in a detached building. Groups H-2 and H-3 cannot be less than 50 feet where a detached building is required.

Occupancies in these two groups containing materials with explosive characteristics must be separated as required by the International Fire Code.

In Groups H-1, H-2, or H-3 where detached buildings are required by Table 4.11 there are no requirements for wall and opening protection based on fire separation distance. There are special provisions for Group H-1 occupancies that you must follow.

Group H-1 occupancies must be in buildings used for no other purpose, cannot exceed one story in height, or be without a basement, crawl spaces, or other under-floor spaces. Roofs must be of lightweight construction with suitable thermal insulation to prevent sensitive material from reaching its decomposition temperature. Group H-1 occupancies containing materials which are in both physical and health hazards in quantities exceeding the maximum allowable quantities per control area must comply with requirements for both H-1 and H-4 occupancies.

Special provisions for Group H-2 and H-3 occupancies containing quantities of hazardous materials must be in buildings used for no other purpose and cannot be more than one story or without a basement or crawl space. If these occupancies contain water-reactive materials they must be resistant to water penetration.

There may not be any piping that carries water placed over or through these areas, unless isolated by approved liquid-tight construction (fire protecting piping is an exception to this). Occupancies in Group H-2 must be constructed in accordance with the following which apply to buildings in which materials that produce combustible dusts are stored. These buildings must comply with NFPA 61, 120, 651, 654, 655, 664, and 85 as well as

Fast Fact

Every room or space occupied for grinding or operations that produce combustible dusts must be enclosed with fire barriers that have no less than a two-hour fire-resistance rating where the area is not more than 3000 square feet, and not less than a four-hour fire-resistance rating where the area is more than 3000 square feet.

the International Fire Code. These buildings must comply with the height and area limitations for Group H-2; except that when erected of Type I or II construction, the heights and areas of grain elevators and similar structures must be unlimited, and where Type IV construction, the maximum height will be 65 feet and in isolated areas, the maximum height of Type IV structures will be increased to 85 feet.

Conveyors

Conveyors, chutes, and piping passing through the enclosures of rooms or spaces must be constructed dirt and vapor tight and be of approved materials complying with Chapter 30. Grain elevators and malt houses cannot be located with 30 feet of inside lot lines or structures on the same lot, except where built along a railroad right-of-way. Coal pockets located less than 30 feet from inside lot lines or from structures on the same lot must be constructed of no less that Type IB construction. If more than 30 feet, the minimum type of construction is not more than 65 feet in height and must be Type IV.

Mixed Occupancies

In mixed occupancies where the storage tank area is located in a building of two or more occupancies, and the amount of liquid is more than allowed, then the use must be completely separated from adjacent fired areas. Any storage tank must be approved and conform to the International Fire Code and must be noncombustible and protected from physical damage. A fire barrier wall must be placed around the tank for protection. A liquid-tight containment area compatible with the stored liquid must be provided, rooms where only double-wall storage tanks conforming to this section are used to store Class I, II, and IIIA flammable liquids are not required to have a leakage container. An approved automatic alarm must be installed to indicate a leak in a storage room or tank. It must be very loud and have the ability to be heard in any room. There will be a sign that states: Warning, When Alarm Sounds, the Environment within the Room May be Hazardous.

All storage tanks need to be vented if they are of Class I or IIIA liquids, this must go to the outdoor air and in accordance to the International Fire Code. All tanks storing Class I, II, or IIIA liquids must be provided with mechanical ventilation. If Class I liquids are being stored, explosion venting is required. The opening of any of these tanks must be designed to ensure that liquid or vapor concentrations are not released into the building.

Liquefied petroleum gases are classified as propane, butane, propylene, and butylenes. Any facilities that house these gases must be provided with air inlets and outlets so that air movement across the floor of the facility will remain even. The total area of both inlet and outlet openings must be at least 1 square inch for each 1 square foot of floor area. The bottom of these openings cannot be more than 6 inches above the floor.

Attached buildings or structures must have a perimeter of no more than 50 percent of the perimeter of the space enclosed. Separation of the attached structures must be provided by fire barriers with a fire-resistance rating of not less than one hour and may not have any openings. Fire barriers be-

Trade Tip

Where located in separate buildings, gas-distribution facilities must be occupied exclusively for that purpose or others of the same nature. These buildings can only be one story high; the floor cannot be located below ground level and must be solidly filled. All construction of each aspect of the building must be constructed of noncombustible materials. Explosion venting must be provided in accordance with the International Fire Code.

Fast Fact

Dry cleaning plants must be constructed in accordance with this code, the International Mechanical Code, the International Plumbing Code, and NFPA 32. Dry cleaning solvents and systems must be classified in accordance to the International Fire Code. When gas rooms are provided, they must be separated from other areas by no less than a one-hour fire barrier.

tween attached structures that are occupied for LP-gas are permitted to have fire door assemblies. These barriers must be designed to withstand a static pressure of at least 100 pounds per square foot (psf), except where the building to which the structure is attached is occupied by operations or processes with similar hazards.

Where liquefied petroleum gas-distribution facilities are located in rooms within buildings, these rooms must be located in the first story level above grade plane, they must also have at least one outside wall with enough exposed area to provide explosion venting. The walls, floors, ceilings, and roofs of such rooms and building must be constructed of approved noncombustible materials, and must be fire barriers with no less than a one-hour fire-resistance rating and without openings. Walls that are used for storing LP-gas are allowed to have openings that comply with code requirements, and they must be designed to withstand a static pressure of at least 100 psf.

Group H-5

Fabrication of Group H-5 that has aggregate quantities of hazardous materials stored and used in a single fabrication must not exceed the amount allowed by this code.

Fabrication areas, whose sizes are limited, must be separated from each other, from corridors, and other parts of the building with one-hour fire barriers, except doors within having a fire protection of not less than ¾ hour. Or, windows between areas and corridors are permitted to be fixed glazing of at least ¾ hour. Occupied levels of areas must be located at or above the first story above grade plane. All floors except for surfacing must be of noncombustible materials. Any shafts, vent shafts, and other openings must be enclosed. The annular space around penetrations for cables, cable trays, tubing, piping, or ducts must be sealed at the floor level to restrict the movement of air.

A ventilation system must be in place to capture and exhaust gases, fumes, and vapors at workstations. Exhaust ducts that penetrate workstations must be contained in a shaft. HPM must be transported to fabrication areas through enclosed piping that comply with this chapter and the International Fire Code.

All electrical equipment used within the fabrication area must conform to the ICC Electrical Code. If the average air change is at least four times an hour and is not less than three times then the location of hazardous materials does not apply to fabrication.

You may not energize workstations without proper exhaust ventilation. All corridors must be separated from fabrication areas as specified in this chapter. Corridors cannot contain HPM and cannot be used for transporting these materials, except through closed piping systems or unless fabrication areas are altered or modified, subject to the following conditions.

Corridors adjacent to the fabrication area where the alteration work is to be done must comply with the code for a length determined as follows: the length of the common wall of the corridor and the fabrication area and the distance along the corridor to the point of entry of HPM into the corridor serving that fabrication area. Also, there must be an emergency telephone system, a local manual alarm station, or another approved alarm-initiating device within the corridors at no more than 150-foot intervals. Or, self-closing doors having a fire protection rating of not less than one hour.

CODE UPDATE

Storage areas for flammable and combustible liquid tanks inside of structures shall be located at or above grade. A two-hour fire barrier is required between the storage area and processing areas.

Service corridors are of a different nature and the coding for them is somewhat different. They are classified as Group H-5 and service corridors are to be kept separate from corridors as previously discussed in this chapters. Service corridors cannot be used as a required corridor and they must be mechanically vented or at no less than six air changes per hour. Service corridors are required to have a means of exit as well as any corridor. The maximum distance of travel from any point to a service corridor is not to exceed 75 feet. Dead ends cannot exceed 4 feet. There may not be less than two exits and no more than one-half can require travel into a fabrication area.

Service corridors must have doors that swing in the direction of the exit. The doors must have a minimum width of 5 feet or 33 inches wider than the widest cart or truck used in the corridor, whichever is greatest. As with required corridors, all service corridors are required to have an emergency alarm system in place.

Storage of hazardous production materials must be within approved or listed storage cabinets or gas cabinets within a workstation. Any HPM in quantities greater than those listed in Section 1804.2 of the International Fire Code must be in liquid storage rooms. The construction of HPM and gas rooms must be separated from other areas and by not less than a two-hour fire barrier in an area that is 300 square feet. Liquid storage rooms must be constructed in accordance with the following:

- Rooms in excess of 500 square feet must have at least one outer door approved for fire department access.
- Rooms must be separated from other areas by fire barriers having a fire-resistance rating of not less than one hour for rooms up to 150 square feet in area.
- Shelving, racks, and wainscoting in these areas must be of noncombustible construction or wood of not less than 1 inch (25 mm) normal thickness.

Rooms used for the storage of Class I flammable liquids must not be located in the basement.

Liquid storage rooms and gas rooms must have a least one outside wall that will be no less than 30 feet from lot lines adjacent to public ways. Explosion control is required. Where two exits are required, one must lead directly outside of the building. Mechanical exhaust ventilation must be provided in liquid storage rooms at the rate of no less than 1 cubic foot per minute of floor area or six air changes per hour. HPM storage is required to have an approved emergency alarm system that is alarm-initiating and sounds a local alarm and transmits a signal to the emergency control station.

Piping and tubing is required of systems that supply HPM liquids or gases having a health-hazard rating of III or IV and must be welded throughout, except for connections to the system that are within a ventilated enclosure. Hazardous production materials supply piping in service must be readily accessible. Where HPM gases or liquids are carried in pressurized piping above 15 pounds per square inch gauge (psig), excess flow control shall be provided. Where the piping comes from within a liquid storage room, HPM room, or gas room, the excess flow control shall be located within the liquid storage room, HPM room, or gas room. The installation of HPM piping and tubing within the space shall be in accordance with the following conditions:

- Automatic sprinklers must be installed within the space unless the space is less than 6 inches in the least dimensions.
- Ventilation not less than six air changes per hour must be provided.
- The space must not be used to convey air from any other area.
- Where the piping or tubing is used to transport HPM liquids, a receptor must be installed below such piping or tubing.
- The receptor must be designed to collect any discharge or leakage and drain it to an approved location.
- The one-hour enclosure must not be used as part of the receptor.
- HPM supply piping and tubing and nonmetallic waste line must be separated from the corridor and from occupancies other than group H-5 by fire barriers that have a fire-resistance rating of not less than one hour.
- Where gypsum wallboard is used, joints on the piping side of the enclosure are not required to be taped, provided the joints occur over framing members.
- Access openings into the enclosure must be protected by approved fire protection-rated assemblies.

- Readily accessible manual or automatic remotely activate fail-safe emergency shutoff valves must be installed on piping and tubing other than waste lines at the following locations: (1) at branch locations into the fabrication area (2) at entries into corridors. The exception to this is any transverse crossing of the corridors by supply piping that is enclosed within a ferrous pipe or tube for the width of the corridor need not comply with bullets one through five.

Gas Detection

Continuous gas detection systems must be provided for HPM gases when the physiological warning threshold level of the gas is at a higher level than the accepted PEL for the gas and for flammable gases in accordance with this section. A continuous gas-detection system is required for fabrication areas, HPM rooms, gas cabinets, and exhausted enclosures and corridors, but is not required for the occasional transverse crossing of the corridors by supply piping that is enclosed in a ferrous pipe or tube for the width of the corridor. The gas-detection system must be capable of monitoring the room, and/or equipment, in which the gas is located at or below the PEL or ceiling limit of the gas. For flammable gases, the monitoring detection threshold level shall be vapor concentrations in excess of 20 percent of the lower explosive limit (LEL). Monitoring for highly toxic gases must also comply with the requirements of the International Fire Code.

Gas detection systems must initiate a local arm and transmit a signal to the emergency control station when a short-term hazard is detected. The gas detection system must automatically close the shutoff valve at the source on gas supply piping and tubing related to the system being monitored for which gas is detected when a short-term hazard is detected. Automatic closure of shutoff valves must comply with the following:

- Where the gas-detection sampling point initiating the gas detection system alarm is within a gas cabinet or exhausted enclosure, the shutoff valve in the gas cabinet or exhausted enclosure for the specific gas detected shall automatically close.
- Where the gas-detection sampling point initiating the gas detection system alarm is within a room and compressed gas

containers are not in gas cabinets or an exhausted enclose, the shutoff valves on all gas lines for the specific gas detected shall automatically close.

- Where the gas-detection sampling point initiating the gas detection system alarm is within a piping distribution manifold enclosure, the shutoff valve supplying the manifold for the compressed gas container of the specific gas detected must automatically close.

The only exception to this is where the gas-detection sampling point initiating the gas detection system alarm is at the use location or within a gas valve enclosure of a branch line downstream of a piping distribution manifold; the shutoff valve for the branch line location in the distribution manifold enclosure must automatically close. The emergency control station must receive signals form emergency equipment and alarm and detection systems. Such emergency equipment and alarm and detection systems must include, but are not be limited to, the following where such equipment or systems are required to be provided, either in this chapter or elsewhere in this code:

- Automatic fire sprinkler system alarm and monitoring systems
- Manual fire alarm systems
- Emergency alarm systems
- Continuous gas-detection systems
- Smoke detection systems
- Automatic detection and alarm systems for phosphoric liquids and Class III water-reactive liquids required in Section 1805.2.3.5 of the International Fire Code
- Exhaust ventilation flow alarm devices for phosphoric liquids and Class III water-reactive liquids required in Section 1805.2.3.5 of the International Fire Code.

Emergency Power

An emergency power system must be provided for Group 5 occupancies. The required electrical systems are to include the following:

- HPM exhaust ventilation systems
- HPM gas cabinet ventilation systems
- HPM exhausted enclosure ventilation systems
- HPM gas room ventilation systems
- HPM gas detection systems
- Emergency alarm systems
- Manual fire alarm systems
- Automatic sprinkler system monitoring and alarm systems
- Automatic alarm and detection systems for phosphoric liquids and Class III water-reactive liquids required in Section 1805.2.3.5 of the International Fire Code
- Flow alarm switches for phosphoric liquids and Class III water-reactive liquids required in Section 1805.2.3.5 of the International Fire Code
- Electrically operated systems required elsewhere in this code or in the International Fire Code applicable to the use, storage, or handling of HPM.

Exhaust Ventilation

Exhaust ventilation systems are allowed to be designed to operate at no less than one-half the normal fan speed of the emergency power system where it is demonstrated that the level of exhaust will maintain a safe atmosphere. Automatic sprinkler system protection in exhaust ducts for HPM must be provided in exhaust ducts conveying gases, vapors, fumes, mists, or dusts generated for HPM. This applies to metallic and noncombustible nonmetallic exhaust ducts where the largest cross-sectional diameter is equal to or greater than 10 inches and the ducts are within the building and they convey flammable gases, vapors, or fumes, except for ducts listed or approved for applications without automatic fire sprinkler system protection or ducts not more than 12 feet in length installed below ceiling level.

OCCUPANCY GROUPS I-1, R-1, R-2, AND R-3

Occupancies in Groups I-1, R-1, R-2, R-3 must comply with provisions of this code. Walls separating dwelling units in the same building and walls

Did You Know?

This section applies to spray rooms where there is spraying of flammable paints, varnishes, and lacquers or other flammable materials, mixtures, or compounds. Spray rooms must be enclosed with fire barriers with no less than one-hour fire-resistance ratings. Floors must be waterproofed and drained in an approved manner. The interior surfaces of spray rooms must be smooth and constructed to permit the free passage of exhaust air. Spraying spaces must be ventilated to prevent the accumulation of flammable mist or vapors.

separating sleeping units in the same building also must comply with this code.

Now, let's move into Chapter 3 to explore more of the mysteries of the building code.

CHAPTER

3

General Building Heights and Areas

Welcome to Chapter 3. In this chapter we will explore general building heights and areas. This chapter includes some definitions that are applied to this code. The purpose of definitions is to keep you updated on words that may take on different meanings in the International Building Code world. I am going to introduce you to a few more groups and their classifications and the codes that these groups must follow.

DEFINITIONS

The following words and definitions applied to this chapter and used in other places in this code have the meanings as described below:

- Area, building: The area included within surrounding exterior walls (or exterior walls and fire walls) exclusive of vent shafts and courts. Areas of the building not provided with surrounding walls must be included in the building area if these areas are included within the horizontal projection of the roof or floor above.

- Basement: The portions of buildings that are partly or completely below grade plane. A basement must be considered as a story above grade plane where the finished surface of the floor above the basement is more than 6 feet above grade plane or more than 12 feet above the finished ground level at any point.
- Equipment platform: An unoccupied, elevated platform used exclusively for mechanical systems or industrial process equipment, including the associated elevated walkways, stairs, and ladders necessary to access the platform.
- Grade plane: A reference plane representing the average of finished ground level adjoining the building at exterior walls. Where the finished ground level slopes away from the exterior walls, the reference plane must be established by the lowest points within the area between the building and the lot line or, where the lot line is more than 6 feet from the building, between the building and a point 6 feet from the building.
- Height, building: The vertical distance from grade plane to the average height of the highest roof surface.
- Height, story: The vertical distance from top to top of two successive finished floor surfaces; and, to the topmost story, from the top of the floor finish to the top of the ceiling joists or, where there is not a ceiling, to the top of the roof rafters.
- Mezzanine: An intermediate level or levels between the floor and ceiling of any story and in accordance with this chapter.

GENERAL HEIGHT AND AREA LIMITATIONS

The height and area for buildings of different construction types will be ruled by the intended use of the building and cannot go over the limits except as modified from now on. Each part of a building included within the inside and outside walls and fire walls, where present, will be allowed to be a separate building. Buildings and structures that are designed to accommodate special industrial processes that require large areas and unusual heights to contain cranes or special machinery and equipment are exempt from the height restrictions, such as:

- Rolling mills

- Structural metal fabrication shops and foundries
- The production and distribution of electricity
- The production and distribution of gas or steam power.

There are situations when two or more buildings are on the same building lot. When this happens, they are to be regulated as separate buildings or they will be considered as parts of one building if the height of each building and the total area of the buildings are within the limits of Table 4.4 as modified by this code. The requirements of this code that are valid to the total building will be appropriate to each building. Buildings that are Type I construction are allowed to be unlimited level heights and areas are not required to stick to the special requirements that allow unlimited area buildings, unlimited height, or increased height and areas for other types of construction in this chapter.

HEIGHT

The height that is allowed by code will be increased in agreement with this section with the exception that the height of one-story aircraft hangars, aircraft paint hangers, and buildings used for the manufacturing of aircraft will not be limited if the building is provided with an automatic fire-extinguishing system and is entirely surrounded by public ways or yards no less in width than one and one-half times the height of the building.

MEZZANINES

Mezzanines that conform to this section can be considered a portion of the story. They cannot, however, be counted as either the building area or the number of stories as regulated by this section. The area of the mezzanine must be included in determining the fire area. The clear floor height of your mezzanine cannot be less than 7 feet.

CODE UPDATE

A basement is a story that is not a story above the grade plane.

Trade Tip

Height includes roof structures. Towers, steeples, and other roof structures must be constructed of materials that are consistent with the required type of construction except where other construction is allowed by Chapter 15. These structures cannot be used for living space or storage. If roof structures, such as towers and steeples, are made of noncombustible materials then their height is unlimited, but they cannot extend more than 20 feet above the allowable height if the structure is not made of noncombustible materials. Chapter 15 has additional information regarding these requirements.

The total area of a mezzanine within a room is not allowed to be over one-third of the floor area of the room or the space that they are in. You also cannot include the enclosed part of the room to determine the floor area where the mezzanine is located. When determining the allowable mezzanine area, the area of the mezzanine cannot be included in the floor area except for the following:

- The total area of mezzanines in buildings and structures that are Type I or II for special industrial occupancies in accordance with this chapter cannot be more than two-thirds of the area in the room.
- The total area of mezzanines in buildings and structures that are Type I or II cannot be more than one-half of the area of the room in buildings and structures that have an approved sprinkler system throughout. The sprinkler system has to be in accordance with code requirements and an approved emergency voice/alarm communication system.

Mezzanines are no different when talking about exits and exit routes. Each occupant of a mezzanine must have access to at least two exits where the common path of exit travel is over the limits of Chapter 10. If the exit from your mezzanine is a stairway, the maximum travel distance must include the distance traveled on the stairway measured in the plane of the tread nosing.

Accessible means of exits must be provided, as well as a single means of exit. If a building or structure has a mezzanine it has to be open and no obstructions are allowed in the room where the mezzanine is located, except for walls that are not more than 42 inches high, columns, and posts.

There are five exceptions to this code, and they are as follows:

- Mezzanines or portions that are of concern are not required to be open, provided that the occupant load does not go over 10 persons.
- Mezzanines or portions that are of concern are not required to be open to the room if at least one of the exits provides direct access to an exit from the mezzanine level.
- Mezzanines are not required to be open to the room, provided that the total floor area of the enclosed space does not go over 10 percent of the area.
- In industrial facilities, mezzanines used for control equipment are allowed to be glazed on all sides.
- In Groups H and I occupancies that are no more than two stories in height above grade plane and equipped with an automatic sprinkler, a mezzanine having two or more exits is not required to be open to the room in which the mezzanine is located.

EQUIPMENT PLATFORMS

Equipment platforms in buildings cannot be considered as a portion of the floor below and must not contribute to either the building area or the number of stories as regulated by this chapter. You may also not use the area of the equipment platform to determine the fire area. Equipment platforms cannot be part of any mezzanine and these platforms and walkways, stairs, and ladders that provide access to an equipment platform cannot be used as an exit from the building either. There are some area limitations that you must be aware of.

The total area of all equipment platforms within a room cannot be larger than two-thirds of the area of the room which they are in. If the equipment platform is located in the same room as a mezzanine, the area of the mezzanine must be determined by this chapter and the combined total area of the equipment platforms and mezzanines cannot be more than two-thirds of the room that they are in. If a mezzanine is in a building that is required to have an automatic sprinkler system, equipment platforms must be fully protected by these sprinklers above and below the platform.

AREA MODIFICATIONS

You know that this code limits the areas that are permitted to be increased with the proper automatic sprinkler system in place. There is an equation that must be applied when increasing these areas in accordance with Equation 3.1.

$$Aa = \{At + \{A1 \times If\} + [A1 \times Is]\} \qquad \text{(Eq. 3.1)}$$

Aa = Allowable area per story in square feet.

At = Tabular area per story in accordance with Table 4.4 in square feet.

If = Area increase factor due to frontage as calculated in accordance with this chapter.

Is = Area increase factor due to sprinkler system protection as calculated in accordance with this chapter.

A single basement that is not a story above plane does need to be included in the total allowable area if it is not bigger than the area that is allowed for a building with no more than one story above grade plane.

Each building must connect or have access to a public way to receive an area increase for frontage. Where a building has more than 25 percent of its outside on a public way or open space that has a minimum width of 20 feet, the frontage increase must be determined in accordance with Equation 3.2.

$$If = [F/P - 0.25]\ W/30 \qquad \text{(Eq. 3.2)}$$

If = Area increase due to frontage.

F = Building perimeter that fronts on a public way or open space having 20 feet open minimum width.

P = Perimeter of entire building in feet.

W = Width of public way or open space (feet) in accordance with this chapter.

Width must be at least 20 feet. Where the value of width varies along the perimeter of the building, the calculation performed in accordance with Equation 5.1 must be based on the weighted average of each part of the exterior wall and open space where the value of "W" is greater than or equal to 20 feet. When "W" is more than 30 feet, a value of 30 feet must be used in calculating the weighted average, regardless of the actual width of the open space. The exception being that the quantity of "W" divided by 30 must be permitted to be a maximum of two when the building meets all requirements of this chapter except for compliance with the 60-foot public way or yard requirement, if applicable. These open spaces must be either on the same lot or dedicated for public use and must be accessed from a street or approved fire lane.

If a building is equipped with an approved automatic sprinkler throughout the building, the area is allowed to be increased by an additional 200 percent (Is = 2) for buildings with more than one story above grade plane and an additional 300 percent (Is = 3) for buildings with no more than one story above grade plane. These increases are permitted in addition to the height and story increases. The following list contains exceptions that must be made to this:

- The automatic sprinkler system increase cannot apply to buildings with an occupancy in Group H-1.
- The automatic sprinkler system increase must not apply to the floor area of occupancy in Group H-2 or H-3. For mixed-use buildings containing these occupancies, the allowable area must be calculated in accordance with this book, with the sprinkler increase applying only to the portions of the buildings not classified as Group H-2 or H-3.

CODE UPDATE

Two or more buildings on the same lot shall be regulated as separate buildings or shall be considered as portions of one building, if the building height of each building and the aggregate building area of the buildings are within code limitations.

- Fire-resistance rating substitution in accordance with code allowances.

The maximum area of a building with more than one story above grade plane must be determined by multiplying the allowable area of the first story (Aa), as determined in the above section, by the number of stories above grade plane as listed below:

- For buildings with three or more stories above grade plane, multiply by three.
- For buildings with two stories above grade plane, multiply by two.
- No story can be larger than the allowable area per story (Aa), as determined in the above section, for the occupancies on that story.

Unlimited area buildings in accordance with this chapter and the maximum area of the building equipped throughout with an automatic sprinkler system determined by multiplying the allowable area per story (Aa), as determined in this chapter, by the number of stories above grade plane are two exceptions to area determination.

In buildings with mixed occupancies, the allowable area must be based on the most restrictive provision for each occupancy when the mixed occupancies are treated in accordance with this chapter. When the occupancies are treated according to this chapter, as separate occupancies, the maximum total building area must be the ratios for each area on all floors as calculated according to this chapter and not go over two for two-story buildings and three for buildings three stories or higher.

UNLIMITED AREA BUILDINGS

The area of buildings and occupancies and configurations of this section will not be limited. The area of a one-story, non-sprinkler Group F-2 or S-2 building must not be limited when the building is surrounded and attached by public ways or yards not less than 60 feet in width. If the area of a one-story, Group B, F, M, or S building or a one-story Group A-4 building, of other than Type V construction, must not be limited when the building is equipped with an automatic sprinkler system throughout, provided it is in accordance with code requirements and it must be sur-

CODE UPDATE

The area of a Group F-2 or S-2 building that is no more than one story in height shall not be limited when the building is surrounded and adjoined by public ways or yards that are not less than 60 feet in width.

rounded and attached by public ways or yards not less than 60 feet in width. Exceptions:

- Buildings and structures of Type I and II construction for rack storage facilities that do not have access by the public must not be limited in height, provided that these buildings conform to the requirements of this code and to NFPA 230.
- The automatic sprinkler system is not required to be in an area occupied for indoor participant sports, such as tennis, skating, swimming, and equestrian activities in occupancies in Group-A, if the exit doors to the outside are provided and the building is equipped with a sprinkler system.
- Group A-1 and A-2 occupancies other than Type V construction must be permitted, provided that all assembly occupancies are separate from other spaces as required with no reduction allowed in the fire-resistance rating and each Group A occupancy cannot be larger than the maximum allowance and all required exits lead directly outdoors.

If a two-story building that is Group B, F, M, or S, and is equipped with an automatic sprinkler system, the area will not be limited. It must, however, be surrounded and adjoined by public ways or yards no less than 60 feet in width. The permanent open space of 60 feet required in this chapter will be permitted to be reduced to no less than 40 feet, if the following requirements are met:

- The reduced open space is not allowed for more than 75 percent of the perimeter of the building.

- The exterior wall facing the reduced open space must have a minimum fire-resistance rating of three hours.
- Openings in the exterior wall facing the reduced open space must have opening protectives with a minimum fire protection rating of three hours.

The area of a one-story, Group A-3 building used as one of the following:

- A place of religious worship
- Community hall
- Dance hall
- Exhibition hall
- Gymnasium
- Lecture hall
- Indoor swimming pool
- Tennis court.

These group uses that must be of Type II construction are not limited when the building does not have a stage other than a platform, if the building is equipped throughout with an automatic sprinkler system, the assembly floor is located at or within 21 inches of street or grade level; all exits provided with ramps and buildings not surrounded and adjoined by public ways or yards no less than 60 feet in width.

Group H occupancies to include H-2, H-3, and H-4 occupancies will be permitted in unlimited area buildings containing Group F and S occupancies, in accordance with this chapter and the limitations of this section. The

Did You Know?

The area of a one-story, Group H-2 aircraft paint hangar will not be limited when it complies with the provisions of this code and is totally surrounded by public ways or yards no less in width than one and one-half times the height of the building.

Fast Fact

Covered mall buildings and anchor stores having an area of no more than three stories in height that comply with this code will not be limited.

total floor area of the Group H occupancies that are located at the outside of the unlimited area building cannot be more than 10 percent of the area of the building or the area limitations for Group H occupancies that are specified in the code as modified by this chapter that are based upon the percentage of the perimeter of each Group H fire area that looks onto on a street or other unoccupied space. Also, the total floor area of Group H occupancies not located at the outside of the building cannot be more than 25 percent of the area limitations for the Group H occupancies. For two-story unlimited area buildings, the Group H fire areas cannot be located above the first story unless permitted by the allowable height in stories and feet as described in this code based on the type of construction of the unlimited area building.

Group E buildings of Type II, IIIA, or IV construction will not be limited when the following criteria are met:

- Each classroom must have no less than two means of exits, with one of those exits being a direct exit to the outside of the building.
- The building is equipped throughout with an automatic sprinkler system.
- The building is surrounded by and joined by public ways or yards no less than 60 feet (18,288 mm) in width.

MIXED USE AND OCCUPANCY

Buildings or parts of buildings that contain two or more occupancies or uses are classified as mixed use. This section applies to mixed use occupancy and the buildings that they occupy. The first subject of this section cover incidental uses that must comply with this section, the only excep-

CODE UPDATE

With some exceptions, no separation is required between accessory occupancies and the main occupancy.

tions are incidental use areas that serve as a dwelling or living space that do not have to comply with this section. An incidental use area must be classified in accordance with the occupancy of that portion of the building in which it is located or the building must be classified as a mixed occupancy and will comply with this section.

Where the code allows an automatic fire-extinguishing system without a fire barrier, the incidental use area must be separated from the rest of the building by construction that is capable of resisting smoke from passing through the building. The partitions must extend from the floor to the underneath of the fire-resistance-rated floor/ceiling assembly or fire-resistance-rated roof/ceiling assembly above or to the bottom of the floor or roof sheathing or sub deck above. Doors must be self-closing or automatic closing when the detection of smoke is made. Doors also must not have any air transfer openings and cannot be undercut in excess of the clearance that is permitted in accordance of NFPA 80.

Where an automatic fire-extinguishing system or automatic sprinkler system is provided, only the incidental use areas need to be equipped with this system.

We are now ready to investigate Chapter 4 to learn more of the meat of the code.

CHAPTER

4

Fire-Resistance-Rated Construction

The definition alert boxes in this chapter contain words that may have a slightly different meaning in the world of building construction. They are located throughout this chapter; look at them, read them, and know them. They are here to help you.

FIRE-RESISTANCE RATING AND FIRE TESTS

The materials that I describe in this chapter for fire resistance will conform to the requirements of this chapter in your code book. The fire-resistance rating of building elements is determined in accordance with test proce-

!Definitionalert

Annular space: The opening around the penetration item.

dures that are made by ASTM E 119 or in accordance with your local code. If you have materials, systems, or devices that have not been tested as part of a fire-resistance-rated assembly, but are incorporated into the assembly, you must show sufficient data or information to the building official that the required fire-resistance rating is not reduced. All of the building materials and methods used in your construction that produce joints and penetration in fire-resistance-rated buildings cannot reduce the required fire-resistance rating except for one condition. When you are determining the fire-resistance rating of outside walls, compliance with the ASTM E 119 data for unexposed surface temperature rise and ignition of cotton waste, because of flame or gases, is only required for the time period that corresponds to the required fire-resistance rating of an outside nonbearing wall that has the same fire separation distance, and must be in a building of the same group.

If inside walls of a building or structure are not proportioned or even, then they have to be tested with both faces of the walls exposed to the furnace and the assigned fire-resistance rating has to be the shortest duration that you get from the two tests that you conduct in agreement with ASTM E 119. If your evidence is given to the building official and shows that the wall was tested with the least fire-resistant side exposed to the furnace, it is the code officer's discretion for approval.

Combustible components such as aggregates are permitted in gypsum and cement concrete mixtures approved for fire-resistance-rated construction. Any approved element or admixture (the product of two substances mixed together) is allowed in assemblies if the tested assembly meets the fire-resistance test requirements. These assemblies that are tested under the

! Definition alert

Fire-resistance rating: The period of time a building element, component, or assembly maintains the ability to confine a fire, continues to perform a given structural function, or both, as determined by the tests, or the methods based on tests, as described in this chapter.

ASTM E 119 cannot be considered restrained unless the evidence that you present is in accordance with ASTM E 119 approved by the building official. The following methods are alternatives for determining fire-resistance rating, but as I said, they must conform to ASTM E 119 and be approved by your building official.

- Fire-resistance designs documented in approved sources
- Prescriptive designs of fire-resistance-rated building elements as approved in this chapter
- Calculations in accordance with this chapter
- Engineering analysis based on comparing building element designs that have fire-resistance ratings determined by ASTM E 119.

There are two types of materials—elementary and composite materials—that are noncombustible; these do not conform to ASTM E 119. Elementary materials that are required to be noncombustible must be tested in accordance with ASTM E 136 and composite materials that have a structural base of noncombustible materials with a surfacing of no more than 0.125 inch thick that have a flame spread index of no more than 50 when tested in accordance with ASTM E 84 will be acceptable as noncombustible materials.

EXTERIOR WALLS

Exterior walls are walls, bearing or nonbearing, used as an enclosing wall, but cannot be a fire wall and have a slope of 60 degrees or greater with the horizontal plane. This section will include moldings, eave overhangs, outside balconies, and similar projections that extend beyond the floor area

!Definitionalert

Horizontal assembly: A fire-resistance-rated floor or roof assembly of materials designed to restrict the spread of fire in which continuity is maintained.

Trade Tip

Projections from walls that are Type III, IV, or V construction can be of any approved material. With the exception of Type V construction, which is allowed for R-3 occupancies, all combustible projections that are located where openings are not allowed or where protection of openings is required must have at least a one-hour fire-resistance-rated construction.

and must comply with this section. Outside balconies and outside stairways that have exits must comply with the code. There are two methods that you must put in place to determine the allowable distance that your walls may jut out. The method that allows for the least amount of distance will be the one that you must use. These methods are: A point one-third the distance to the lot line from an assumed vertical plane located where protected openings are required in accordance with this chapter or more than 12 inches into areas where openings are prohibited. If the building or structural wall projections are from Type I or Type II construction they have to be made of materials that are noncombustible or combustible as allowed by code exceptions.

OUTSIDE WALLS

Outside walls have protected and unprotected openings. These openings are required to be vertically separated so that they are protected against fire spread on the outside of the buildings if the openings are within 5 feet of each other horizontally, and the opening in the lower story is not a protected opening with a fire protection rating of no less than ¾ hour. But, you must also have the openings vertically separated at least 3 feet by spandrel girders or outside walls that have a one hour fire-resistance-rating. If not, you have to install flame barriers that extend at least 30 inches horizontally beyond the outside wall. Please note that whether you choose spandrel girders or flame barriers, a one hour fire-resistance rating is required. Exceptions to this are as follows:

- This section does not apply to buildings that are three stories or less in height.
- This section does not apply to buildings that are fully equipped with an automatic sprinkler system, but it must be throughout the whole building.
- This section does not apply to parking garages.

For buildings that are located on the same lot, opening protectives that have a fire-protection rating that is less than ¾ hour must be in every opening that is less than 15 feet vertically above the roof of the building that is next to it within 15 feet of the wall where the opening is located. But if the roof construction has a fire-resistance rating no less than one hour with a minimum distance less than 10 feet then you can disregard the above. The International Building Code requires that all outside walls include parapets, which are walls placed at the edge of a roof. There are certain conditions in which a parapet does not need to be added to an outside wall:

- The wall is not required to be fire-resistance-rated because of fire separation distance.
- The building's area is not more than 1000 square feet on any floor.
- Walls that end at roofs that are less than two-hour fire-resistance construction or if the roof, this includes the deck and supports, is made entirely of noncombustible materials.
- One-hour fire-resistance walls that end at the underside of the roof, deck, or slab, if the following needs are met:
  - Roof and ceiling frames cannot be less than one-hour fire-resistance with a width of 4 feet for Groups R and U and 10 feet for all other occupancies.
  - Roof and ceiling elements that are not parallel to the wall, the entire span of framing and supports cannot be less than one-hour fire-resistance construction.
  - Openings in the roof cannot be located within 5 feet of any one-hour fire-resistant outside wall for Groups R and U and 10 feet for all other occupancies.
  - The entire building must be provided with a roof covering that is no less than Class B.

- Groups R-2 and R-3 that are covered with a Class C roof covering must have outside walls that end at the underside of the roof sheathing in Type III, IV, and V construction provided that the following conditions are met:
  - The roof or deck is constructed of noncombustible materials that are approved or made of fire-retardant wood with a distance of 4 feet.
  - The roof is protected with 0.625-inch Type X gypsum board directly beneath the roof or deck. The supports of the roof must have a minimum of 2-inch ledgers attached to the sides of the roof framing with a distance of 4 feet.
  - The wall is allowed to have at least 25 percent of the outside walls containing unprotected openings.

PARAPET CONSTRUCTION

Parapets must have the same fire-resistant rating as other construction of any outside wall. They also must be noncombustible from 18 inches from the top downward (this includes counterflashing and coping materials). The code also requires that the parapet is not less than 30 inches above the point of the roof and wall intersection. And, the parapet must extend to the same height of the roof within fire separation distance, but in no case should the height be more than 30 inches. With any rule or code in the International Code Book, if you have any questions, please consult a building official.

Did You Know?

Any windows in outside walls that are required to have protected openings must comply with this chapter; other openings have to be protected with fire doors. Opening protectives are not required if the building is protected throughout with automatic sprinkler systems or approved water curtains, and joints that are made in outside walls are allowed to have unprotected openings.

CODE UPDATE

Fireblocking consists of materials installed to resist the free passage of flame to other areas of the building through concealed spaces.

FIRE WALLS

Each portion of a building that is separated by one or more fire walls must be considered as a separate building. The extent and the location of the fire walls have to provide a complete separation. Wherever a fire wall separates occupancies you must follow the strictest requirements. This includes any wall located on a lot line between buildings that are adjacent to each other or is adapted for joint service, and must be constructed as a fire wall. Fire walls have to be constructed with the same fire conditions as the rest of the building in regards to duration of time. All fire walls, with the exception of Type V construction, have to be made of approved noncombustible materials.

Fire walls must be continuous from outside to inside walls and must extend at least 18 inches beyond the outside surface walls. Exceptions to this rule are found below:

- Fire walls are permitted to end at the inside surface of combustible outside siding provided that the outside walls have a fire-resistance rating of at least one hour for a horizontal distance of at least 4 feet on both sides of the fire wall. Any openings with outside walls have to be protected by opening protectives that have a rating not less than ¾ hour.
- Fire walls are permitted to end at the inside surface of a noncombustible outside casing where the buildings on each side of the fire wall are protected by an automatic sprinkler system.

Exterior or outside walls, at the point of intersection, must have, on both sides, a one-hour fire rating with ¾ hour opening protection. The rating must extend a minimum of 4 feet on each side starting from the intersection to the outside all. If such walls form an angle equal to or greater than

180 degrees they do not need outside wall protection. Fire walls must extend to the outer edge of horizontal projections such as balconies, roof overhangs, canopies, and other similar projects that are within 4 feet of the fire wall. There are a few exceptions in regards to these projections that I will explain in the following list:

- Horizontal projections without concealed spaces, provided that the outside wall behind and below the projection has no less than a one-hour fire rating for a distance that is not less than the depth of the projection on either side of the fire wall. These openings with outside walls have to be protected by openings with a fire protection rating of ¾ hour; nothing less.
- Noncombustible horizontal projections with concealed spaces providing a minimum one-hour rated construction, extending through the concealed space. The wall is not required to extend under the projection and the openings within the outside walls must be protected by opening protectives that have a fire rating of ¾ hour; nothing less.
- For combustible horizontal projections with concealed spaces, the fire wall only needs to extend through the concealed space to the outer edges of the projections, and again, a one hour fire rating of ¾ hour is required.

Fire walls must have vertical continuity, meaning fire walls have to extend from the foundation to an ending point at least 30 inches above both adjacent roofs. However, two hour fire-rated walls will be permitted to end at the underside of the roof, deck, or slab if the lower roof within 4 feet of the wall has a one-hour fire rating, openings in the roof are not located within 4 feet of the fire wall, and each building has no less than a Class B roof covering.

CODE UPDATE

Fire walls, fire barriers, fire partitions, smoke barriers, and smoke partitions are required to be permanently identified with signs or stenciling.

Fast Fact

When a fire wall serves as an outside wall for a building and separated buildings that have different roof levels, the wall must end at a point no less than 30 inches above the lower roof level and the outside wall has to be at least 15 feet above the lower roof. These must have a one-hour fire rating construction from both sides of the openings protected by fire assemblies which must be protected by no less than a ¾ hour rating.

Walls will be permitted to end at the underside of noncombustible roofs, decks, or slabs provided that both buildings have no less than a Class B roof covering. In buildings of Type III, IV, and V construction, walls are allowed to end at the underside of the roof or deck if there are no openings in the roof within 4 feet of the fire wall, the roof is covered with a minimum Class B roof covering, and the roof is constructed of fire-retardant treated wood with a distance of 4 feet on both sides of the wall or the roof is protected with Type X gypsum that is ⅝ inch thick. If the building is located above a parking garage it will be allowed to have fire walls for the buildings located above the garage from the horizontal separation between the parking garage and the buildings.

If a fire wall ends at the base of a roof, the lower roof within 10 feet of the wall has no less than a one-hour fire rating, and the entire length and span of the supports for the roof assembly cannot have a rating of less than one-hour. The lower roof cannot have openings that are located within 10 feet of the fire wall.

Each opening through a fire wall must not be more than 120 square feet, unless both buildings have an automatic sprinkler system throughout. The total width of openings at any floor level cannot be over 25 percent of the length of the wall and openings are not allowed in party walls. Penetrations of fire walls and joints made in or between fire walls must comply with this chapter, and remember; ducts and air transfer openings cannot penetrate fire walls unless they are not on a lot line and comply with this chapter.

FIRE BARRIERS

Fire barriers must be made of materials that are permitted by the building type of construction and fire-resistance-rated glazing, when tested in accordance with ASTM E 119 shall comply with this section. They have to be labeled or show some type of identification, provided by an approved agency that has the name of the manufacturer, the test standard and the identifier that includes the fire-resistance rating. Exit enclosures, exit passageways, to include horizontal exits, all must comply with the code book.

When outside walls are part of a required fire-rated shaft or exit enclosure they have to be in accordance with the code. Fire barriers must extend from the top of the floor/ceiling assembly below to the bottom of the floor or roof slab or deck above and be securely attached.

Fire barriers must also be continuous through concealed spaces, such as the space above a suspended ceiling. All supporting construction must be protected to allow the required fire rating of the fire barrier supports. Hollow vertical spaces within a fire barrier must be fire blocked at every floor level except that shaft enclosures are allowed to end at a top enclosure when complying with this section.

All outside walls that are used as a part of a required fire-rating enclosure or separation have to comply with this section except for outside walls that are in accordance with code exceptions. Openings in a fire barrier will be protected and will be limited to a maximum aggregate width of 25 percent of the length of the wall having a maximum area of any single opening not being more than 156 square feet. This does apply to openings that have ad-

Did You Know?

Fire windows that are permitted in atrium separation walls are not limited to the maximum width of 25 percent. All penetrations of fire barriers must comply with this section and penetrations into an exit enclosure or passageway will be allowed, but only when permitted by the code.

CODE UPDATE

King studs and boundary elements that are integral elements in load-bearing walls of light-frame construction shall be permitted to have required fire-resistance ratings provided by the membrane production provided for the load-bearing wall.

joining fire sprinklers throughout and the opening assembly has been tested with ASTM E 119, fire doors serving as an exit.

SMOKE BARRIERS

All smoke barriers are required to have a one-hour fire-resistance rating with the exception of smoke barriers that are a minimum of 0.10-inch-thick steel in Group I-3 buildings. Smoke barriers must form an effective casing that is continuous from outside wall to outside wall and from the top of the foundation or floor/ceiling assembly below to the bottom of the floor or roof above. This is to include the connection through concealed spaces, such as those found above suspended ceilings and mechanical spaces. Any supporting construction has to be protected to allow the required fire-resistance rating of the wall or floor supported in buildings other than Type IIB, IIIB, or VB construction. Any smoke barrier with penetrations, joints, and ducts and air transfer openings must comply with this chapter.

SMOKE PARTITIONS

Smoke partitions are not required to have a fire-resistance-rating unless required to do so elsewhere in this book. Smoke partitions extend from the top of the foundation or floor to the bottom of the roof to limit the transfer of smoke. All windows and doors, but not to include louvers, must be tested in accordance with UL 1784 before they are considered to be able to resist the free passage of smoke. Any free spaces around penetrating items and in joints must be filled with an approved material to limit the free passage of smoke. This is also true for ducts and air transfer openings, unless installation will interfere with the operation of a required smoke control system.

HORIZONTAL ASSEMBLIES

A horizontal assembly refers to roofs and floors. Floor and roof assemblies that are required to have a fire-resistance rating must comply with this section. All floor and roof assemblies must be made of materials allowed by the type of building construction and the fire-resistance rating of floor and roof assemblies cannot be less than that required by the building type of construction. But what if you have floor assemblies that separate mixed occupancies? In cases such as this, the assembly must have a fire-resistance rating of no less than the minimum required by code, based on the occupancies that are being separated.

If a floor assembly separates a single occupancy into different fire areas, the assembly must have a fire-resistance rating of no less than what was discussed earlier in this chapter. All floor assemblies that separate living spaces in the same building or sleeping units in occupancies in Group R-1, hotel occupancies, R-2, and I-1 must be a minimum of one-hour fire-resistance-rated construction. The only exception to this is dwelling and sleeping units that are used as separation in buildings of Type IIB, IIIB, and VB construction which must have a fire-resistance rating of no less than ½ hour in buildings that are equipped throughout with an automatic sprinkler system.

Ceiling Panels

Where the weight of lay-in ceiling panels that are used as part of fire-resistance-rated floor/ceiling or roof/ceiling assemblies is not strong enough to resist an upward force of 1 lb (48 Pa), you must install wire or another approved device to prevent vertical bowing under such upward force. Access doors are permitted in ceilings of fire-resistance-rated assemblies provided that the doors are tested in accordance with ASTM E 119 as horizontal as-

CODE UPDATE

Fire barriers shall extend from the top of the floor/ceiling assembly below to the underside of the floor or roof sheathing, slab, or deck above and shall be securely attached thereto.

semblies and labeled as such. In one-hour-fire-resistance-rated floor construction and one-hour fire-resistance-rated roof construction, neither the ceiling covering nor the floor covering is required to be installed where unusable crawl space or unusable attic space occurs. All assemblies must be continuous without openings, penetrations, or joints except where it is allowed elsewhere in this code.

PENETRATIONS

Where sleeves are used, they must be securely fastened to the assembly that is penetrated. The space between the item contained in the sleeve and the sleeve itself and any space between the sleeve and the assembly penetrated must be protected. Insulation and coverings on or in the penetrating item must not penetrate the assembly unless the specific material used has been tested as part of the assembly.

Penetration into or through fire walls, fire-barrier walls, smoke-barrier walls, and fire partitions must comply with the following:

- In concrete or masonry walls where the penetrating item is a maximum of 6-inch nominal diameter and the area of the opening through the wall is not more than 144 square inches, concrete, grout, or mortar is allowed where it is installed or the full thickness required to maintain the fire-resistance rating or if the material used to fill the annular space will prevent the passage of flame and hot gases that can ignite cotton waste when put up against ASTM E 119 time-temperature fire conditions under a minimum positive pressure difference of 0.01 inch of water at the location of the penetration for the time period that is equal to the fire-resistance rating of the construction penetrated.
- All penetrations must be installed as tested in an approved fire-resistance rated assembly.
- Penetrations must be protected by an approved firestop system installed and tested in accordance with ASTM E 814 or UL 1479, with a minimum positive pressure differential of 0.01 inch of water and must have an F rating of no less than required of the fire-resistance rating.

Trade Tip

Dissimilar materials such as noncombustible penetrating items must not connect to combustible items beyond the point of firestopping unless it can be demonstrated that the fire-resistance integrity of the wall is maintained.

- Where walls or partitions are required to have a fire-resistance rating, recessed fixtures cannot be installed in a way that will reduce the required fire resistance, except for the following: Membrane penetrations of a maximum two-hour fire-resistance-rated wall and partition by steel electrical boxes that are not more than 16 square inches in area, but only if the total area of the openings through the membrane is not more than 100 square inches in any 100 square feet of wall area. The annular space between the wall membranes and the box cannot be more than 1/8 inch.

Steel electrical boxes on opposite sides of walls or partitions must be separated by one of the following:

- By a horizontal distance of no less than 24 inches
- By a horizontal distance of not less than the depth of the hole in the wall where the hole is filled with cellulose loose-fill, rockwool, or slag mineral wool insulation
- By solid fireblocking
- By protecting both outlet boxes with listed putty pads
- By any other listed materials and methods
- Membrane penetrations by listed electrical boxes of any material, and the boxes have been tested for use in fire-resistance-rated assemblies
- Boxes are installed in accordance with the instructions that are included in the listing

- The space between the wall and the box cannot be more than ⅛ inch unless listed otherwise
- Boxes on opposite sides of the wall or partition must be separated according to the list above.

FIRE-RESISTANCE JOINT SYSTEMS

Joints that are installed in or between fire-resistance-rated walls, floors, or floor/ceiling assemblies, and roofs or roof/ceiling assemblies must be protected by an approved fire-resistant joint system designed to resist the passage of fire for a time period that is not less than the required rating. The void or space created at the intersection of a floor/ceiling assembly and an outside curtain wall assembly must be protected in accordance with this section. Fire-resistant joint systems will not be required for joints in all of the following locations:

- Floors within a single dwelling unit
- Floors where the joint is protected by a shaft enclosure
- Floors within atriums where the space adjacent to the atrium is included in the volume of the atrium for smoke control purposes
- Floors within malls
- Floors within open parking structures
- Mezzanine floors
- Walls that are allowed to have unprotected openings
- Roofs where openings are permitted
- Control joints that are not more than a maximum width of 0.625 inch (15.9 mm) and tested in accordance with ASTM E 119.

Fire-resistant joint systems must be securely installed in or on the joint for its entire length so it will not dislodge, loosen, or otherwise impair its ability to accommodate expected building movements and to resist the passage of fire and hot gases. Fire-resistant joint systems must be tested in accordance with the requirements of either ASTM E 1966 or UL 2079. Non-symmetrical wall joint systems must be tested with both faces exposed to the furnace, and the assigned fire-resistance rating must be the shortest time from the two tests.

CODE UPDATE

A separate fan system shall be used for each elevator hoistway.

When evidence is furnished to show that a wall was tested with the least fire-resistant side exposed to the furnace, and the building official agrees, then the wall does not need to be subjected to tests from the opposite side. For outside walls with a horizontal fire separation distance more than 5 feet, the joint system must be required to be tested for inside fire exposure only.

Where fire-resistance-rated floor or floor/ceiling assemblies are required, empty spaces created at the intersection of the outside curtain wall assemblies and such floor assemblies must be sealed with an approved material or system to prevent fire from spreading around the inside. These materials or systems must be securely installed and be capable of preventing the passage of flames and hot gases that are hot enough to ignite cotton waste when subjected either to ASTM E 119 time-temperature fire conditions under a minimum positive pressure differential of 0.01 inch of water column or installed as tested in accordance with ASTM E 2307 for the time period at least equal to the fire-resistance-rating of the floor assembly.

Height and fire-resistance requirements for curtain wall spandrels must comply with the section of this chapter that does not require a fire-resistance-rated spandrel wall. Fire-resistant joint systems in smoke barriers must be tested in accordance with the requirements of UL 2079 for air leakage. The air leakage rate of the joint must not be more than 5 cfm per lineal foot of joint at 0.30 inch of water for both the immediate surrounding area temperature and elevated temperature tests.

FIRE-RESISTANCE RATING OF STRUCTURAL MEMBERS

The fire-resistance rating of structural members and assemblies must comply with the requirements for the type of construction and cannot be less than the rating required for the fire-resistance-rated assemblies supported except for fire barriers and partitions and smoke barriers that were provided earlier in this chapter. Protection of columns, girders, trusses, beams, lintels, or other structural pieces that are required to have a fire-resistance rating must also comply with this section. Any of these that are required to have a fire-resistance rating and that support more than two floors or one floor and roof, or support a load-bearing wall or a non-load-bearing wall more than two stories high, must be individually protected on all sides for the full length with materials having the required fire-resistance rating.

King studs and boundary elements that are integral elements in load-bearing walls of light-framed construction must be permitted to have required fire-resistance ratings provided by the covered protection provided for the load-bearing wall. Where columns need a fire-resistance rating, the entire column, including its connections to beams or girders, must be protected.

Where a column extends through a ceiling, fire resistance of the column must be continuous from the top of the foundation or floor/ceiling assembly below through the ceiling space to the top of the column. The required thickness and construction of fire-resistance-rated assemblies enclosing trusses must be based on the results of full-scale tests or combinations of tests on truss components or on approved calculations based on such tests that satisfactorily demonstrate that the assembly has the required fire resistance. The edges of lugs, brackets, rivets and bolt heads that are attached to structural members must be allowed to go over one inch of the surface of the fire protection.

The thickness of protection for concrete or masonry reinforcement must be measured to the outside of the reinforcement except that stirrups and spiral reinforcement ties are allowed to project no more than 0.5 inch into the protection. Any and all pipes, wires, conduits, ducts, or other service facilities must not be embedded in the required fire protective covering of a structural member that is required to be individually encased.

Any place where the fire protective covering of a structural member is subject to impact damage from moving vehicles, the handling of merchandise or other

CODE UPDATE

Penetrations in smoke barriers must be tested for air leakage.

CODE UPDATE

A listed smoke damper designed to resist the passage of smoke shall be provided at each point a duct or air transfer opening penetrates a fire wall that serves as a horizontal exit.

activity, the fire protective covering must be protected by corner guards or by a substantial jacket of metal or other noncombustible material to a height that will provide full protection, but no less than 5 feet from the finished floor.

Exterior Structural Members

Exterior structural members that are located within the outside walls or on the outside of a building or structure must be provided with the highest fire-resistance rating as determined in accordance for the type of building element based on the type of construction of the building, for outside bearing walls and as required for outside walls based on the fire separation distance.

Fire protection is not required at the bottom flange of lintels, shelf angles, and plates spanning no more than 6 feet whether part of the structural frame or not, and from the bottom flange and not part of the structural frame regardless of span.

Fire-resistance ratings for the isolation system must meet the fire-resistance rating required for columns, walls, or other structural elements in which the isolation system is installed in accordance with code requirements. Isolation systems that are required to have a fire-resistance rating must be protected with approved materials or construction assemblies designed to provide the same degree of fire resistance as the structural element in which it is installed when tested in accordance with ASTM E 119. With this isolation protection system applied to isolator units it must be capable of preventing the transfer of heat to the isolation unit in such a manner that the required gravity load-carrying capacity of the isolator unit will not be damaged after exposure to the standard time-temperature curve fire test as in ASTM E 119 for no less than required for the fire-resistance rating of the structure element of installation. This isolation system protection applied to isolator units must be correctly designed and securely installed so it does not dislodge, loosen, become damaged, or harm its ability to accommodate the seismic movements for which it is designed.

OPENING PROTECTIVES

Any opening protectives that are required by other sections of this code must comply with this section. However, labeled fire-resistance-rated glazing tested as part of a wall assembly in accordance with ASTM E 119 does not need to comply with this section. There are alternative methods for determining fire protection ratings, but they must be based on the fire exposure and acceptance criteria specified in NFPA 252 or NFPA 257. The required fire resistance of an opening protective must be permitted to be recognized by any of these methods:

- Designs documented in approved sources
- Calculations performed in an approved manner
- Engineering analysis based on a comparison of opening protective designs having fire-protection ratings as determined by the test procedures as found in NFPA 252 or 257
- Alternate protection methods.

Doors

I talked earlier about fire doors and shutter assemblies and the fire ratings for such. These doors do not include side-hinged or pivoted swinging doors that must be tested as well in accordance with NFPA 252 or UL 10C.

Did You Know?

Doors that are in exit enclosures and exit passageways must have a maximum transmitted temperature end point of no higher than 450°F above the surrounding area at the end of 30 minutes of a standard fire test exposure, but the maximum transmitted temperature rise is not limited in buildings equipped throughout with an automatic sprinkler system.

This test requires that after five minutes of testing with the NFPA 252 test, the neutral pressure level in the furnace must be at 40 inches (1016 mm) or less above the sill.

There are other types of doors, such as swinging elevator doors, that also must be tested in accordance with NFPA 252 or UL 10B. The pressure in the furnace must be maintained as equal as possible to the atmospheric pressure. Once this is done, the pressure must be maintained during the entire test period. Fire door assemblies that are required to have a minimum fire protection rating of 20 minutes where located in corridor walls or smoke-barrier walls having a fire-resistance rating in accordance with code must be tested in accordance with NFPA 252 or UL 10C without the hose stream test. Listed below are three exceptions to this:

- Corridor door assemblies in occupancies of Group I-2 must be in accordance with code regulations.
- Unprotected openings must be permitted for corridors in multi-theater complexes where each auditorium has as least one-half of its required exits or exit access doorways opening directly to the exterior or into an exit passageway.

Fire door assemblies must meet the requirements for a smoke and draft control door assembly tested in accordance with UL 1784. Louvers are prohibited. When you are installing smoke doors be sure to follow NFPA 105. In a 20-minute fire door assembly, the glazing material in the door itself must have a minimum fire-protection rating of 20 minutes and must be exempt from the hose stream test. Glazing material in any other part of the door assembly, including transom lites and sidelites, must be tested in accordance with NFPA 257, and includes the hose stream test.

All fire door assemblies, oversized doors, smoke and draft doors must be labeled by an approved agency and must comply with NFPA 80. I will remind you that these types of labels have to be permanently attached to the door or the frame. The label must also show the name of the manufacturer, the name of the inspection agency, the fire rating and, if required, the maximum temperature end point.

Some of you may be using wired glass panels that are found in some fire doors. The codes for these types of doors differ slightly from the ones I mentioned above. Glazing other than wired glass in fire windows must be fire-protection-rated glazing and installed in accordance with NFPA 80. It must be installed in the fixed position or be automatic-closing.

Fire-protection-rated glazing in fire doors that are located in fire walls are prohibited except where used as a horizontal exit. A self-closing swinging door will be permitted to have a vision panel of no more than 100 square inches without a dimension more than 10 inches and they cannot be installed in fire doors that have a 1½ -hour fire protection rating intended for installation in fire barriers, unless the glazing is not more than 100 square inches in area.

All approved fire-protection-rated glazing that is used in fire door assemblies in elevator and exit enclosures must be located sot that there is a clear view of the hallway or entry to the elevator, ramp, or stairway. These too must be labeled as stated above. Fire-protection-rated labels must have one of the following four-part identification: "D–H," "NH–T," or "NT–XXX." The list below provides the meaning of each of these.

- "D" indicates that the glazing must be used in fire door assemblies and it meets the fire resistance of the test.
- "H" indicates that the glazing meets the hose stream requirements of the test standards.
- "NH" indicates that the glazing does not meet the hose stream requirements.
- "T" indicates that the glazing meets the temperature requirements.

Trade Tip

All fire dampers used for hazardous exhaust duct systems must comply with the International Mechanical Code. Assemblies with less than three-hour fire-resistance-rating must have a minimum of 1.5 hour damper rating. And assemblies that have a three or more hour rating must have a minimum of three hour damper rating.

- "NT" indicates that the glazing does not meet the temperature requirements.
- "XXX" specifies the fire-protection-rating period, in minutes.

Fire protection rated glazing that is installed in fire doors or fire window assemblies in hazardous locations must comply with the code. Fire doors that are located in common walls that separate sleeping areas in Group R-1 are not required to be automatic or self-closing; always check with your local building official as this is not the rule for all fire doors, and as a reminder, automatic-closing doors must be in accordance with NFPA 80. Below is a list of all doors that must be automatic closing. Also, unless otherwise permitted, all fire doors, either single or side-hinged swinging doors, must have a latch bolt that secures the doors when closed. Glazing in fire window assemblies must comply with the code in regards to fire protection ratings.

- Doors installed across a corridor
- Doors that protect openings in exits
- Doors that protect openings in walls and are capable of resisting smoke
- Doors installed in smoke barriers
- Doors installed in fire partitions

Fast Fact

Smoke damper leakage ratings cannot be less than Class II and elevated temperature ratings cannot be less than 250°F. Smoke dampers must close when a smoke detector(s) is activated. In places where a damper is installed within a duct, a smoke detector must also be installed within 5 feet of the damper with no air outlets or inlets in between. If a damper is installed above a smoke barrier door in a smoke barrier, a spot-type detector listed for releasing service must be installed on either side of the smoke barrier door opening. If a damper is installed within an unducted opening in a wall, a spot-type detector listed for releasing service must be installed within 5 feet horizontally of the damper.

- Doors installed in a fire wall
- Doors installed in shaft enclosures
- Doors installed in refuse and laundry chutes and access and termination rooms
- Doors installed in the walls for compartmentation of underground buildings
- Doors that are installed in elevator lobby walls of underground buildings
- Doors that are installed in smoke partitions.

If your building or structure has installed vertical sliding or vertical rolling steel fire doors for pedestrian use, they must be heat activated or activated by smoke detectors to ensure the safety of the public.

FIRE DAMPERS

Ducts and air transfer openings that penetrate fire partitions must be protected with listed fire dampers. Occupancies other than Group H do not require a fire partition if the following are met:

- The buildings are tenant separation or corridor walls in buildings that are equipped throughout with an automatic sprinkler system and the duct is protected as a through penetration
- Tenant partitions in covered mall buildings where the walls are not required by provisions elsewhere in the code to extend to the bottom of the floor or roof deck above
- The duct system is constructed of approved materials in accordance with the International Mechanical Code and the duct that penetrates the wall meets the following requirements:
  - The duct cannot be more than 100 square inches.
  - The duct must be constructed of steel that is a minimum of 0.0217 inch thick.
  - The duct cannot have openings that share the corridor with adjacent spaces or rooms.
  - The duct cannot terminate as a wall register in the fire-resistance-rated wall.

- A minimum 12-inch-long by 0.060-inch-thick sleeve has to be centered in each duct opening.
- The sleeve will be secured to both sides of the wall and all four sides of the sleeve must have steel retaining angles that are a minimum of 1.5 inches by 1.5 inches by 0.060 inch.
- The retaining angles must be secured to the sleeve and the wall with no. 10 screws.
- The annular space between the steel sleeve and the wall opening must be filled with mineral wool batting on all sides.

Smoke Damper

A listed smoke damper is designed to resist smoke passing through areas of buildings and is required to be provided at each point that a duct or air transfer opening penetrates a corridor. However please note that smoke dampers are not required when the building has an approved smoke control system in place throughout and they are also not required in corridor penetrations if the duct is constructed of steel that is no less than 0.019 inch thick and there are no openings in use for corridors.

You must also have a smoke damper in place at each duct or air transfer opening penetrating a smoke barrier unless the openings in the duct is a single smoke compartment and constructed of steel. Penetrations by ducts and air transfer openings of a floor or floor/ceiling assemblies or the ceil-

Did You Know?

Ducts and air transfer openings constructed of approved materials in accordance with the International Mechanical Code that penetrate the ceiling membrane of a fire-resistance-rated floor/ceiling or roof/ceiling assembly must be protected in a shaft enclosure or a listed ceiling radiation damper that is installed at the ceiling line where the duct penetrates the ceiling of floor/ceiling or roof/ceiling or a ceiling radiation damper that is installed at the ceiling line where a diffuser with no duct attached penetrates the ceiling of a floor/ceiling or roof/ceiling assembly.

CODE UPDATE

Sprayed fire-resistant materials shall be identified by density and thickness required for a given fire-resistance rating.

ing covering of a roof/ceiling assembly must be protected by a shaft enclosure that complies with this chapter. In occupancies other than Groups I-2 and I-3, a duct that is constructed of approved materials in accordance with the International Mechanical Code and penetrates a fire-resistance-rated floor/ceiling that connects no more than two stories is permitted without shaft enclosure protection, granted that a listed fire damper is installed at the floor line or the duct is protected.

A duct is permitted to penetrate three floors or less without a fire damper at each floor, provided it meets all of the following requirements:

- The duct must be contained and located within the cavity of a wall and must be constructed of steel that is no less than 0.019 inch in thickness.
- The duct must open into only one dwelling or sleeping unit and the duct system must be continuous from the unit to the outside of the building.
- The duct cannot be more than 4 inches nominal diameter and the total area of the ducts cannot be more than 100 square inches in any 100 square feet of floor area.
- The annular space around the duct is protected with materials that prevent the passage of flame and hot gases sufficient to ignite cotton waste when exposed to ASTM E 119 time-temperature conditions under a minimum positive pressure difference of 0.01 inch of water at the location of the penetration for the time period equal to the fire-resistance rating of the construction penetrated.
- Grilled openings located in a ceiling of a fire-resistance-rated floor/ceiling or roof/ceiling assembly must be protected with an installed listed ceiling radiation damper.

CONCEALED SPACES

In combustible concealed spaces fire-blocking and draft-stopping must be installed. Fire-blocking materials must include nominal or standard lumber that is at least 2 inches thick or two pieces of 1 inch lumber with broken lap joints. Other materials can include: one wood structural panel of at least 0.719 inch that has joints backed by another wood structural panel of the same thickness. You may also use gypsum board, cement fiber board, batts or blankets of mineral wool, glass fiber, or any other approved materials. Always check with the building officer or your local code office to obtain approval of your materials in question.

All materials must be securely installed if to be considered for an appropriate fire-block. Loose-fill insulation material cannot be used unless specifically tested in the form and manner intended for use and shown to be an appropriate material that will remain in place and will repel the spread of fire and hot gases.

In walls that are made of double stud, batts and blankets of mineral or glass fiber will be allowed as fire-blocking in walls using parallel rows of studs or staggered studs. Speaking of stud walls and partitions, which include furred spaces and parallel rows of studs, they must be vertically at the ceiling and floor levels or horizontally at intervals that are not more than 10 feet.

Fire-blocking must be provided for the following areas:

- Connections between horizontal and vertical spaces that are between the stud walls
- Concealed horizontal spaces created by an assembly of floor joists or trusses
- Between concealed horizontal and vertical spaces such as soffits and drop and cove ceilings

!Definitionalert

Sleeper: Usually, a wood member embedded in concrete, as in a floor, that serves to support and to fasten the sub-floor or flooring.

- In concealed spaces in stairways between stringers at the top and the bottom of the run
- Annular spaces in openings around vents, pipes, ducts, chimneys and fireplaces at ceiling and floor levels. Note: factory-built chimneys and fireplaces must be fire-blocked in accordance with UL 103 and UL 127.

Architectural trim that has concealed spaces of outside wall finish and other outside architectural elements must also have fire-blocking installed where erected with combustible frames, at maximum intervals of 20 feet, so that there are no open spaces more than 100 square feet. If wood furring strips are used, they need to be made of an approved wood that has a natural decay resistance or preservative-treated wood.

Fire-blocking is not required of cornices in single-family homes, however in two-family homes you must install fire-blocking, but only at the line of separation between the two homes. If the framing of the building or structure and the face of the outside wall finish is covered by aluminum that is at least 0.019 inch thick or corrosion-resistant steel with a base metal thickness at least 0.016 inch or any other approved materials, you are not required to install fire-blocking.

In concealed sleeper spaces where wood sleepers are used for laying wood floors on masonry or concrete fire-resistance-rated floors, the space between the floor slab and the bottom of the wood flooring must be filled with an approved material to resist the passage of flame and products of combustion of fire-blocked so that there will be no open spaces under the flooring that will be more than 100 square feet in area.

Any empty spaces must be filled so that they are solid full under permanent partitions and there is no passing under the flooring between adjacent rooms. This includes the joint at each alternate lane and at the ends of each lane in a bowling alley. However, gymnasium floors that are slab-on-grade are not required to have fire-blocking.

Draft Stopping

Fire-blocking has a "friend" that we must not forget about—draft-stopping. One cannot be without the other. All combustible construction must have draft-stopping installed to sub-divide floor/ceiling assemblies in locations and occupancies that follow. There are several approved materials that you can use for draft-stopping. Any material that you use cannot be less than 0.5 inches. But first, let's take a look at the required materials that draft-stopping must be made of:

- Gypsum board—0.375 inch (9.5 mm)
- Wood structural panel—0.375 inch (9.5 mm)
- Particle board—1 inch (25-mm), nominal lumber
- Cement fiberboard, batts, or blankets of mineral wool or glass fiber, or other approved materials that offer enough support can also be used for draft-stopping.

Now that you know what materials you can use for draft-stopping, I will go into the occupancies in which draft-stopping must in installed. Groups R-1, R-2, R-3, and R-4 all require draft-stopping in certain situations.

Group R-1 must provide draft-stopping in floor/ceiling spaces. Draft-stopping must be provided in Group R-2 buildings with three or more dwelling units, Group R-3 buildings with two dwelling units, and in Group R-4 buildings. Draft-stopping must be located above and in line with the dwelling unit and sleeping unit separations. Draft-stopping must also be installed in attics to sub-divide attic and concealed roof spaces. Draft-stopping is not required in buildings equipped throughout with automatic sprinkler systems as long as the sprinklers are also installed in the combustible concealed spaces. In all other groups, draft-stopping must be installed so that horizontal floor areas are not more than 1000 square feet.

In Groups R-1 and R-2 draft-stopping must be provided in attics, mansards, overhangs, or other concealed roof spaces of Group R-2 buildings with three or more dwelling units in all Group R-1 buildings. Draft-stopping must be installed above, and in line with, sleeping unit and dwelling unit separation walls that do not extend to the bottom of the roof sheathing above. There are three exceptions however that you must be aware of:

- In cases where corridor walls provide a sleeping unit or dwelling unit separation, draft-stopping is only required above one of the corridor walls.

- Draft-stopping is not required in buildings that are equipped throughout with an automatic sprinkler system, only if combustible concealed spaces are included in the system.
- In occupancies in Group R-2 that are not more than four stories in height, the attic space does not need to be subdivided by draft-stops into an area that is not more than 3000 square feet or above every two dwelling units, or whichever is smaller.

THERMAL AND SOUND-INSULATING MATERIALS

Insulating materials that include facings such as vapor retardants and vapor-permeable membranes, similar coverings, and all layers of single and multilayer reflective foil insulations must comply with this section. Flame spread indexes or a smoke-developed index must be determined with ASTM E 84. Any material that is subject to an increase in flame spread index or smoke-developed index beyond the limits of this code because of the age of the material, moisture, or other conditions caused by exposure to the atmosphere for a long period of time are not permitted.

Concealed and exposed insulation when installed in buildings of any type of construction must always have a flame index of no more than 25 and smoke-developed index of no more than 450. This does not apply to buildings of Type III, IV, or V construction regarding the flame spread for facings, coverings, and layers of reflective foil insulation that are installed behind and in contact with the unexposed surface of the ceiling, wall, or floor finish. However, exposed insulation that is installed on attic floors must have a critical radiant flux of no less than 0.12 watt per square centimeter when tested in accordance with ASTM E 970.

Loose-Fill Insulation

Another type of insulation is loose-fill. This type of insulation cannot be heaped in the ASTM E 84 apparatus without a screen or support unless it complies with the flame spread and smoke-developed limits. If the loose-fill insulation that you are using is cellulose it does not have to comply with the frame spread index that is required of CAN/ULC S102.2. But, you need to be aware that loose-fill insulation does have to comply with CPSC 16 CFR, Part 1209, and CPSC 16, Part 1404 when applicable and must be labeled as such. Combustible roof insulation that does not comply with this section can be used in any type of construction, but it needs to be covered with an approved roof covering.

We have completed this chapter and are now ready to move into Chapter 5.

CHAPTER

5

Fire Protection Systems

This chapter specifies where fire protection systems are required and how they must apply to the design, installation, and operation of fire protection systems. I have included definitions exclusive to this chapter subject. Installation, repair, operation, and maintenance of fire protection systems must be in accordance with this code and the International Fire Code. This includes modifications which are required to follow this code and the International Fire Code; you must also get approval from the building official—there are no exceptions to this rule.

If you are using or installed a fire protection system or part of one that is not required by this code, it also must meet the requirements of this code. When you are installing threads that are provided for fire department connections to sprinkler systems, standpipes, yard hydrants, or any other fire hose connections you must make certain that they are compatible with the connections that are used by the local fire department. A simple phone call or visit to the fire department can save time and money or even a code decision by ensuring that you've made sure that these connections are a match.

All fire protection systems must be tested in accordance with the requirements of this code and the International Fire Code. If testing is required it is your responsibility that it is conducted in the presence of the code official. It is not enough to have the test(s) done; the building official must see the tests from start to finish. The costs of these tests are at the expense of the owner or a person representing the owner. It is very important that you realize that it is illegal to occupy a structure or even a portion of the structure if the proper testing has not been done and/or approved by the building official.

Fire protection systems must be monitored (where required) by an approved supervising station in accordance with NFPA 72. All automatic sprinkler systems must also be monitored by an approved supervising station with the exception of automatic sprinkler systems protecting one- and two-family housing units and limited area systems that serve fewer than 20 sprinklers.

Fire alarm systems require similar monitoring with an approved supervising station with the exception of single- and multiple-station smoke alarms and smoke detectors in Group I-3 occupancies. Manual fire alarms, automatic fire-extinguishing, and emergency alarm systems in Group H occupancies must be monitored by an approved supervising station as well.

When approved by the building official, on-site monitoring at a constantly attended location can be permitted. This is true only if the owner provides that notifications to the fire department will be equal to those provided by an approved supervising station.

DEFINITIONS

Below you will find definitions that pertain to this chapter; I have also scattered them throughout the chapter. These words and terms, for the purpose of this chapter, will have the meanings shown.

!Definitionalert

Alarm notification appliance: A fire alarm system component such as a bell, horn, speaker, light, or text display that provides audible, tactile, or visible outputs or any combination of these.

!Definitionalert

Alarm verification feature: A feature of automatic fire detection and alarm systems to reduce unwanted alarms where smoke detectors report alarm conditions for a minimum period of time, or confirm alarm conditions within a given time period, after being automatically reset, in order to be accepted as a valid alarm-initiation signal.

AUTOMATIC SPRINKLER SYSTEMS

All automatic sprinkler systems installed in a structure must follow the rules or codes of this section. If you plan on or are using alternative automatic fire-extinguishing systems complying with this chapter, they must be recognized by the applicable standard and approved by the fire code official and all new buildings and structures must be provided with approved automatic sprinkler systems and comply with this code. This chapter covers classifications of buildings by group and informs you of what the code specifies for these group classifications.

Group A

Group A requires an automatic sprinkler system for Group A-1 occupancies where one or more of the following conditions exist:

- The fire area exceeds 12,000 square feet.
- The fire area has an occupant load of more than 300 people.

!Definitionalert

Audible alarm notification appliance: A notification appliance that alerts by the sense of hearing.

- The fire area is located on a floor other than the level of exit doors.
- The fire area contains a multi-theater complex.

The following list pertains to Group A-2 occupancies and the following conditions in which automatic sprinkler systems are mandatory by this code:

- The fire area exceeds 5000 square feet.
- The fire area has an occupant load of 100 or more.
- The fire area is located on a floor other than the level of exit doors.

Group A-3 and Group A-4

An automatic sprinkler system must be provided for Group A-3 and Group A-4 occupancies as the following conditions exist:

- The fire area exceeds 12,000 square feet.
- The fire area has an occupant load of 300 or more.
- The fire area is located on a floor other that the level of exit doors.

Group A-3 and A-4 have the same exception to these rules: any areas used exclusively as participant sport areas where the main floor area is located at the same level as the level of exit of the main entrance and exit.

Group A-5

Group A-5 occupancies must be equipped with an automatic sprinkler system in the following areas:

CODE UPDATE

A heat detector is a fire detector that senses heat that has an abnormally high temperature or a rate of rise, or both.

- Concession stands
- Retail areas
- Press boxes
- Other accessory use areas in excess of 1000 square feet.

Group E

Group E occupancies require automatic sprinklers throughout all areas that are greater than 20,000 square feet in area and throughout every portion below the level of exit discharge of buildings used for educational purposes. The exception being that an automatic sprinkler system is not required in any fire area or area below the level of exit discharge where every classroom throughout the building has at least one exterior exit door at ground level.

Group F-1

In Group F-1 occupancies automatic sprinkler systems must be provided throughout all buildings where one of the following conditions exists:

- Where a Group F-1 fire area is more than 12,000 square feet
- Where a Group F-1 fire area is located more than three stories above grade plane
- Where the combined area of all Group F-1 fire areas on all floors, including any mezzanines, is more than 24,000 square feet.

Also Group F-1 occupancies such as those which contain woodworking operations in excess of 2500 square feet in area that cause finely divided combustible waste or use finely divided combustible materials are required to have automatic sprinkler systems installed.

Group H

High hazard occupancies, such as Group H, must have an automatic sprinkler installed. The design of the sprinkler system cannot be less than that required by this code for the occupancy hazard classification. Where the design area of the sprinkler system consists of a corridor protected by one

Trade Tip

All Group R buildings are required to provide automatic sprinklers throughout.

row of sprinklers, the maximum number of sprinklers required to be calculated is 13. Also keep in mind that buildings in which cellulose nitrate film or pyroxylin plastics are manufactured, stored, or handled in quantities of more than 100 pounds are required to be equipped with automatic sprinklers.

Group I

Institutional groups which are classified as Group I are required to install automatic sprinkler systems. Any automatic sprinkler systems that are of an NFPA 13R sprinkler system and installed in accordance to code are allowed to be installed in Group-I institutional groups. Group M occupancies which have one of the following conditions are required to have an automatic sprinkler system installed throughout the building. These conditions are:

- Any Group M fire area that is more than 12,000 square feet
- A Group M fire area that is located more than three stories above grade plane
- Where the combined area of all Group M fire areas on all floors, including any mezzanines, exceeds 24,000 square feet.

Did You Know?

Buildings of Group M occupancies in which storage of merchandise is piled high or put into rack storage arrays are required to have an automatic sprinkler system which must be in accordance with the International Fire Code.

!Definitionalert

Constant attended location: A location designated at a facility that is staffed on a continuous basis by trained personnel. Alarms and supervisory signals are monitored and facilities are provided for notification of the fire department or other emergency services.

Group S-1

In Groups S-1 the entire building is required to be equipped with an automatic sprinkler system where one of the following conditions exists:

- Any Group S-1 fire area that is more than 12,000 square feet.
- A Group S-1 fire area is located more than three stories above grade plane.
- The combined area of all Group S-1 fire areas on all floors, including any mezzanines, is more than 24,000 square feet.

MONITORING AND ALARMS

When speaking of fire safety and automatic sprinkler systems we must look at sprinkler system monitoring and alarms. All valves controlling the water supply for automatic sprinkler systems, pumps, tanks, water levels and temperatures, critical air pressures, and water-flow switches on all sprinkler systems must be electrically supervised. Exceptions to this rule are:

- Automatic sprinkler systems protecting one-and two-family dwellings
- Limited area systems serving fewer than 20 sprinklers
- Automatic sprinkler systems installed in accordance with NFPA 13R where a common supply main is used to supply both domestic water and the automatic sprinkler systems and a separate shutoff valve for the automatic sprinkler system is not provided

- Jockey pump control valves that are sealed or locked in the open position
- Control valves to commercial kitchen hoods, paint spray booths, or dip tanks that are sealed or locked in the open position
- Valves controlling the fuel supply to fire pump engines that are sealed or locked in the open position
- Trim valves to pressure switches in dry, pre-action, and deluge sprinkler systems that are sealed or locked in the open position.

Alarm, supervisory, and trouble signals must be distinctly different and automatically sent to an approved central station, remote supervising station, or proprietary supervising station as defined in NFPA 72 or, when approved by the fire code official, must sound an audible signal at a constantly attended location. There are two exceptions to this rule. The first exception states that underground key or hub valves in roadway boxes provided by the municipality or public utility are not required to be monitored.

The second exception states that backflow prevention device test valves located in limited area sprinkler system supply piping must be locked in the open position. In occupancies required to be equipped with a fire alarm system, the backflow preventer valves must be electrically supervised by a tamper switch installed in accordance with NFPA 72 and separately annunciated.

Alarm devices must be provided on the outside of the building in an approved location. Where a fire alarm system is installed, the motion of the automatic sprinkler system must set off the building fire alarm system. Don't forget this very important detail: all sprinkler systems must be tested and maintained in accordance with the International Fire Code.

Fast Fact

Alarms, such as audible devices, must be approved and connected to every automatic sprinkler system. These sprinkler water-flow alarm devices must be activated by water flow equivalent to the flow of a single sprinkler of the smallest orifice size installed in the system.

CODE UPDATE

A notification zone is an area within a building or facility covered by notification appliances which are activated simultaneously.

ALTERNATIVE SYSTEMS

If you plan on installing an alternative automatic fire-extinguishing system there are codes that also need to be obeyed. Any automatic fire-extinguishing systems, other than automatic sprinkler systems, must be designed, installed, inspected, tested, and maintained in accordance with the next section and the applicable referenced standards. Any automatic fire-extinguishing systems installed as an alternative to the required automatic sprinkler systems in this chapter must be approved by the fire code official. You may not use automatic fire-extinguishing systems as alternatives for the purpose of exceptions or reductions that are allowed by other requirements of this code. What this means is that a building owner cannot substitute with an alternate system in hopes of applying this to an exception to the code. It just doesn't work in this manner.

Before conducting final acceptance tests, the following items must be inspected:

- Hazard specification for consistency with design hazard
- Type, location, and spacing of automatic-and manual-initiating devices
- Size, placement and position of nozzles or discharge holes
- Location and identification of audible and visible alarm devices
- Identification of devices with proper designations
- Operating instructions.

All notification appliances, connections to fire alarm systems, and connections to approved supervising stations must be tested in accordance with this chapter to verify proper operation. Also keep in mind that the audibil-

Did You Know?

If you have installed wet-chemical, dry-chemical extinguishing systems, they must be installed, maintained, and periodically inspected and tested in accordance with NFPA 17 and their listing. If you have installed foam-extinguishing systems they must be installed, maintained, periodically inspected, and tested in accordance with NFPA 11 and NFPA 16 and their listing.

ity and visibility of notification appliances signaling agent discharge or system operation, where required, must be verified. These too need to be tested to verify proper identification and re-transmission of alarms from automatic fire-extinguishing systems.

COMMERCIAL COOKING

Occupancies that have commercial cooking systems must be of a type recognized for protection of commercial cooking equipment and exhaust systems of the type and arrangement protected. Pre-engineered automatic dry- and wet-chemical extinguishing systems must be tested in accordance with UL 300 and listed and labeled for the intended application.

Automatic fire-extinguishing systems of the following types must be installed in accordance with the referenced standard indicated, as follows:

- Carbon dioxide extinguishing systems, NFPA 12
- Automatic sprinkler systems, NFPA 13
- Foam-water sprinkler system or foam-water spray systems, NFPA 16
- Dry-chemical extinguishing systems, NFPA 17
- Wet-chemical extinguishing systems, NFPA 17A.

An exception to this code is factory-built commercial cooking re-circulating systems that are tested in accordance with UL 710B and listed, labeled, and installed in accordance with Section 304.1 of the International Mechanical Code.

!Definitionalert

Standpipe system: An arrangement of piping valves, hose connections, and allied equipment installed in a building or structure with the hose connections located in such a manner that water can be discharged in streams or spray patterns through attached hoses and nozzles.

Commercial cooking systems must have a manual actuation device to be located at or near a means of exit from the cooking area a minimum of 10 feet and a maximum of 20 feet from the kitchen exhaust system. The manual actuation device must be installed not more than 48 inches or less than 42 inches above the floor and must clearly identify the hazard protected. The manual motion must require a maximum force of 40 pounds and a maximum movement of 14 inches to set the fire suppression system in motion.

Automatic sprinkler systems are not required to be equipped with manual movement means. This fire suppression system must automatically shut down the fuel or electrical power supply to the cooking equipment and the supply must be re-set manually. If and when carbon dioxide systems are used, there must be a nozzle at the top of the ventilating duct. Additional nozzles that are balanced to give uniform distribution must be installed within vertical ducts larger than 20 feet and horizontal ducts larger than 50 feet.

Dampers must be installed at either the top or the bottom of the duct and must be arranged to operate automatically upon activation of the fire-extinguishing system. If the damper is installed at the top of the duct, the top nozzle must be immediately below the damper.

CODE UPDATE

Sprinkler protection is required for exterior balconies, decks, and ground floor patios of dwelling units where the building is of Type V construction, provided there is a roof or deck above.

The sizing of an automatic carbon dioxide fire-extinguishing system is extremely important to protect against all hazards venting through a common duct at the same time. If your cooking equipment is a commercial type, it must be arranged to shut off the ventilation system upon activation.

STANDPIPE SYSTEMS

Standpipe systems must be provided in new buildings and structures in accordance with this chapter. Fire hose threads used in connection with standpipe systems must be approved and must be compatible with fire department hose threads. The location of fire department hose connections must be approved as well. When it comes to standpipe systems, there are many code rules that must be followed.

All standpipe systems must be installed in accordance with this section and NFPA 14. Standpipe systems are allowed to be combined with automatic sprinkler systems, but are not required in Group R-3 occupancies.

There are various classes of standpipes to be aware of when it comes to installation of standpipes. Class III standpipe systems must be installed throughout buildings where the floor level of the highest story is located more than 30 feet above the lowest level of fire department vehicle access, or where the floor level of the lowest story is located more than 30 feet below the highest level of fire department vehicle access. There are exceptions to this rule.

CLASS I

Class I standpipes have many conditions associated with their use. The list below will begin our investigation into Class I systems:

- Class I standpipes are allowed in buildings equipped throughout with an automatic sprinkler system in accordance with this chapter.

> **CODE UPDATE**
>
> Cabinets that are used to house portable fire extinguishers shall not be locked.

CODE UPDATE

Portable fire extinguishers are not allowed to be concealed from view, unless suitable signage exists to show the location of the fire extinguishers.

- Class I manual standpipes are allowed in open parking garages where the highest floor is located not more than 150 feet above the lowest level of fire department vehicle access.
- Class I manual dry standpipes are allowed in open parking garages that are subject to freezing temperatures, provided that the hose connections are located as required for Class II standpipes in accordance with this section.
- Class I standpipes are allowed in basements equipped throughout with an automatic sprinkler system.
- In determining the lowest level of fire department vehicle access, it must not be required to consider recessed loading docks for four vehicles or less or conditions where topography makes access from the fire department vehicle to the building impractical or impossible.

Class I automatic wet standpipes must be provided in non-sprinklered Group A buildings having an occupant load exceeding 1000 persons except for open-air-seating spaces that are without enclosed spaces or Class I automatic dry and semi-automatic dry standpipes or manual wet standpipes are allowed in buildings where the highest floor surface used for human occupancy is 75 feet or less above the lowest level of fire department vehicle access.

Some groups have special requirements when it comes to standpipes such as covered mall buildings. The code clearly states that a building such as this must be equipped throughout with a standpipe system where required by this chapter.

As the owner or contractor of a building you must provide the following hose connections in the locations that it states:

- Within the mall at the entrance to each exit passageway or corridor
- At each floor-level landing within enclosed stairways opening directly on the mall
- At exterior public entrances to the mall.

Stages are another classification that requires standpipes. Any stage that is greater than 1000 square felt in area must be equipped with a Class III wet standpipe system with 1.5 inch and 2.5 inch hose connections on each side of the stage. The exception to this is where the building or area is equipped throughout with an automatic sprinkler system; the 1.5 inch hose connection must be installed in accordance with NFPA 13 or in accordance with NFPA 14 for Class II or III standpipes. The 1.5 inch hose connections must be equipped with sufficient lengths of 1.5 inch hose to provide fire protection for the stage area.

Hose connections must be equipped with an approved adjustable fog nozzle and be mounted in a cabinet or on a rack. Make yourself aware that all underground buildings must be equipped throughout with a Class I automatic wet or manual wet standpipe system.

Buildings that are equipped with a helistop or heliport that contain a standpipe must extend the standpipe to the roof level on which the helistop or heliport is located in accordance with Section 1107.5 of the International Fire Code. Marinas and boatyards must also be equipped throughout with standpipe systems in accordance with NFPA 303. The following list contains all locations that Class I standpipe hose connections must be provided:

- In every required stairway, a hose connection must be provided for each floor level above or below grade.
- Hose connections must be located at an intermediate floor level landing between floors, unless otherwise approved by the fire code official.
- On each side of the wall adjacent to the exit opening of a horizontal exit, except where floor areas that are adjacent to a horizontal exit are reachable from exit stairway hose connections by a 30-foot hose stream from a nozzle attached to 100

Trade Tip

All occupancies and locations must be equipped with a portable fire extinguisher as required by the International Fire Code.

feet of hose, a hose connection must not be required at the horizontal exit.

- In every exit passageway at the entrance from the exit passageway to other areas of a building
- In covered mall buildings, adjacent to each exterior public entrance to the mall and adjacent to each entrance from an exit passageway or exit corridor to the mall
- Where the roof has a slope less than four units vertical in 12 units horizontal, each standpipe must be provided with a hose connection located either on the roof or at the highest landing of stairways with stair access to the roof. An additional hose connection must be provided at the top of the most hydraulically remote standpipe for testing purposes.
- Where the most remote portion of a nonsprinklered floor or story is more than 200 feet from a hose connection, the fire code official is authorized to require that additional hose connections be provided in approved locations.

CLASS II AND CLASS III STANDPIPE SYSTEMS

Class II standpipe hose connections must be accessible and located so that all portions of the buildings are within 30 feet of a nozzle attached to 100

CODE UPDATE

Smoke detectors installed in ducts shall be listed for the air, velocity, temperature, and humidity present in the duct.

feet of hose. Group A-1 and A-2 occupancies with occupant loads of more than 1000 hose connections must be located on each side of any stage, on each side of the rear of the auditorium, on each side of the balcony, and on each tier of dressing rooms. Fire-resistance-rated protection of riser and laterals of Class II standpipe systems is not required, but for Class II systems a minimum 1-inch hose is permitted to be used for hose stations in light-hazard occupancies where investigated and listed for this service and where approved by the fire code official. Location of Class III standpipe hose connections has the same requirements of those for Class I standpipes and must have Class II hose connections as required in this chapter. You must not forget that risers and laterals of Class III standpipe systems must be protected as required for Class I systems as well. In buildings where more than one Class III standpipe is provided, the standpipes must be interconnected at the bottom.

As with all code regulations you must contact your local building code official for any additional information or for any questions you may have. There is no such thing as a stupid question. This code is designed to protect everyone, so please be sure that you understand the code and question the code when needed.

ALARM AND DETECTION SYSTEMS

Construction documents for fire alarm systems must be permitted for review and approval before you install any system. Construction documents must include, but are not be limited to, all of the following:

- A floor plan which indicates the use of all rooms
- Locations of alarm-initiating and notification appliances
- Alarm control and trouble signaling equipment
- Annunciation
- Power connection
- Battery calculation
- Conductor type and sizes
- Voltage drop calculation
- Manufacturers, model numbers, and listing information for equipment, devices, and materials

Did You Know?

Don't misinterpret this code for contents, restorations of operations, or for assistance in fire suppression or overhaul activities. Smoke control systems that are regulated by this section serve a different purpose than the smoke-and heat-venting provisions that I will get into later. Also keep in mind that mechanical smoke control systems must not be considered exhaust systems under the International Mechanical Code.

- Details of ceiling height and construction
- The interface of fire safety control functions.

Systems and their components must be listed and approved for the purpose for which they are installed. An approved manual, automatic, or manual and automatic fire alarm system installed in accordance with the provisions of this code and NFPA 72 must be provided in new buildings and structures in accordance with this chapter, where applicable, and provide occupant notification unless other requirements are provided by another chapter or section of this code.

Automatic sprinkler protection that is installed in accordance with this section is provided and connected to the building fire alarm system; automatic heat detection required by this section is not required. The automatic fire detectors must be smoke detectors. Where ambient conditions prohibit installation of automatic smoke detection, other automatic fire detection must be allowed.

SMOKE CONTROL

You must install and maintain smoke control systems. The simple issue with smoke control systems is to be able to control smoke during a fire incident so that the danger is contained and the building can be safely evacuated. This section applies to mechanical or passive smoke control systems when they are required by other provisions of this code. The purpose of this section is to establish the minimum requirements for the design, installation, and acceptance testing of smoke control systems that are intended to provide a tenable environment for the evacuation or relocation of occupants.

Buildings, structures, or parts thereof are required by this code to have a smoke control system(s) designed in accordance with the applicable requirements of this section and the generally accepted and well-established principles of engineering relevant to the design. All construction documents must include sufficient information and detail to adequately describe the elements of the design necessary for the proper implementation of the smoke control systems. These documents must be accompanied by sufficient information and analysis to demonstrate compliance with these provisions.

All buildings and structures are required to have ordinary inspections and test requirements, but smoke control systems are mandated to undergo special inspections and tests sufficient to verify the proper commissioning of the smoke control design in its final installed condition. You must include detail procedures and methods to be used and the items subject to these inspections and tests on the design submission documents. This commissioning must be in accordance with generally accepted engineering practice and, where possible, based on published standards for the particular testing that is involved.

The smoke control system must be supplied with two sources of power. Primary power must be from the normal building power system. Secondary power must be from an approved standby source complying with the ICC Electrical Code.

CODE UPDATE

Where delayed egress locks are installed on means of egress doors an automatic smoke or heat detection system shall also be installed.

The standby power source and its transfer switches must be in a separate room from the normal power transformers and switch gear. It must be enclosed in a room that is constructed of not less than one-hour fire barriers that is ventilated directly to and from the outside. All power distribution from the two sources must come from independent routes. The transfer to full standby power has to be automatic and within 60 seconds of failure of the primary power. This system is required to comply with this code or the ICC Electrical Code. It is imperative that control systems for mechanical smoke control systems include provisions for verification. A verification list follows:

- Positive confirmation of actuation
- Testing
- Manual override
- The presence of power downstream of all disconnects
- A preprogrammed weekly test sequence
- Report abnormal conditions audibly, visually, and by printed report.

Activation of smoke control systems must follow code regulations. Mechanical smoke control systems that use pressurization, airflow, or exhaust method must have completely automatic control. Passive smoke control systems actuated by approved spot-type detectors listed for releasing service are permitted. Where your smoke control system is completely automatic, the automatic-control sequence has to be initiated from an appropriately zoned automatic sprinkler system complying with this section, manual controls that are readily accessible to the fire department and any smoke detectors required by engineering analysis.

Don't forget that any detection and control systems must be clearly marked at all junctions, accesses, and terminations and identical control diagrams showing all devices in the system and identifying their location and function must be maintained current and kept on file with the fire code official, the fire department, and in the fire command center in a format and manner that has been approved by the fire chief.

A fire-fighter's smoke control panel for fire department emergency response purposes only must be provided and has to include manual control

CODE UPDATE

The height of the lowest horizontal surface of the smoke layer interface shall be maintained at least 6 feet above any walking surface that forms a portion of a required egress system within the smoke zones.

or override of automatic control for mechanical smoke control systems. The panel will be located in a fire command center in high-rise buildings or buildings with smoke-protected assembly seating. In all other buildings, the fire-fighter's smoke control panel must be installed in an approved location adjacent to the fire alarm control panel. The fire-fighter's smoke control panel must comply with the International Building Code book.

With all these fire and smoke alarm systems in place we cannot forget smoke and heat vents. Where smoke and heat vents are required by this code, or otherwise installed, smoke and heat vents, or mechanical smoke exhaust systems, and draft curtains must conform to the requirements of this section. There are two exceptions to this. Frozen food warehouses used solely for storage of Class I and II merchandise where protected by an approved automatic sprinkler system or areas of buildings that are equipped with early suppression fast response (ESFR) sprinklers, automatic smoke and heat vents are not required.

Smoke and heat vents must be installed in the roofs of Group F-1 and S-1 buildings that have occupancy or more than 50,000 square feet, except for Group S-1 aircraft repair hangars.

CHAPTER

6

Means of Egress

The provisions of this chapter control the design, construction, and arrangement of means of egress, or simply put, exit tools that are required to provide an approved means of exit from structures or portions thereof. It is against code to alter a building or structure in a manner that will reduce the number of exits or the capacity of the exits to be any less than is required by this code. All means of exits must be maintained in accordance with the International Fire Code.

GENERAL MEANS OF EXITS

The general requirements specified in this section apply to all three elements of the means of exit, or a way out or egress (please note that the words way out, exit, and egress may be used interchangeably and have the same meaning) system, in addition to those specific requirements for the exit access, the exit, and the exit discharge detailed elsewhere in this chapter.

While at the ceiling construction phase keep in mind there must be a minimum of 80 inches for headroom for any walking surface, including walks, corridors, aisles, and passageways. Not more than 50 percent of the ceiling area of the means of exit can be reduced in height by protruding objects; an exception to this is door closers and stoppers. The code states that these cannot reduce headroom to less than 78 inches and you must provide a barrier where the vertical clearance is less than 80 inches high.

The leading edge of any barrier must be located 27 inches (maximum) from the floor. If you have installed a free-standing object that is mounted on a post or pylon, it cannot overhang more than 4 inches where the lowest point of the leading edge is more than 27 inches and less than 80 inches above the walking surface.

If a sign or other obstruction is mounted between posts or pylons and the clear distance between the posts/pylons is greater than 12 inches then the lowest edge of the sign or obstruction must be 27 inches maximum or 80 inches minimum above the finished floor or ground, except that this requirement shall not apply to sloping portions of handrails serving stairs and ramps.

Structural elements, fixtures, or furnishings cannot project horizontally from either side more than 4 inches over any walking surface between the heights of 27 inches and 80 inches above the walking surface (handrails that service stairs and ramps are permitted to protrude 4.5 inches from the wall). Protruding objects will not reduce the minimum clear width of accessible routes.

Note that all walking surfaces of the means of exits must have a slip-resistant surface and be securely attached. This will protect occupants when traveling from rooms to exits. In those places where changes in elevation of less than 12 inches exist in the way of exits, sloped surfaces must be used. If the slope is greater than one side or the other, you must have a ramp in-

CODE UPDATE

An aisle is an unenclosed exit access component that defines and provides a path of egress travel.

CODE UPDATE

A flight is a continuous run of rectangular treads, winders or combination thereof from one landing to another.

stalled that complies with this chapter. If the difference in elevation is 6 inches or less, the ramp must be equipped with either handrails or floor finish materials that contrast with adjacent floor finish materials. The following is a list of exceptions that pertain to the elevation change:

- A single step with a maximum riser height of 7 inches is permitted for buildings with occupancies in Groups F, H, R-2, and R-3 and Groups S and U at outside doors not required to be accessible.
- A stair with a single riser or with two risers and a tread is permitted at locations not required to be accessible provided that the risers and threads comply with this chapter, have a minimum tread depth of 13 inches, and at least one handrail that complies with this chapter is provided within 30 inches of the centerline of the normal path of exit travel on the stair.
- A step is allowed in aisles serving seating that has a difference in elevation less than 12 inches at locations not required to be accessible provided that the risers and treads comply with this chapter and the aisle has a handrail complying with this chapter.

Any change in elevation in a corridor serving non-ambulatory persons in Group I-2 occupancy must be by means of a ramp or sloped walkway. The path of the exit cannot be interrupted by any building element other than a means of egress component as specified in this chapter. Obstructions cannot be placed in the required width of a means of exit except projections that are permitted by this chapter. The required capacity of a means of exit system cannot be diminished along the path of exit travel. You cannot count elevators, escalators, and moving walks as a component of a required means of egress from any other part of the building, except when elevators are used as an accessible means of egress in accordance with this chapter.

OCCUPANT LOAD

When you are constructing a building, you must always use the number of occupants when determining any means of egress or exit. In areas where occupants from accessory areas exit through a primary space, the calculated occupant load for the primary space must include the total occupant load. Every room or space that is an assembly occupancy must have the occupant load of the room or space posted in a conspicuous place, near the main exit or exit access doorway from the room or space. Posted signs must be of an approved legible permanent design and must be maintained by the owner or authorized agent.

Where exits serve more than one floor, only the occupant load of each floor considered individually must be used in computing the required capacity of the exits at that floor, provided that the exit capacity will not decrease in the direction of exit travel. Where means of egress from floors above and below converge at an intermediate level, the capacity of the means of egress from the point of junction must be less than the sum of the two floors.

The code contains the maximum floor area allowances per occupant. The number of occupants must be computed at the rate of one occupant per unit of area. Where an intended use is not listed, the building official will establish a use based on a listed use that most nearly resembles the intended use. The exception being that where approved by the building official, the actual number of occupants for whom each occupied space, floor, or building is designed, although less than those determined by calculation will be permitted to be used in the determination of the design occupant load.

For areas that have fixed seats and aisles, the occupant load must be determined by the number of fixed seats installed. The occupant load for areas in which fixed seating is not installed, such as waiting areas and wheelchair spaces, must be determined in accordance with this chapter and added to

CODE UPDATE

Protruding objects shall not reduce the minimum clear width of accessible routes.

the number of fixed seats. For areas that have fixed seating without dividing arms, the occupant load cannot be less than the number of seats bases on one person for each 18 inches of seating length. The occupant load of seating booths must be based on one person for each 24 inches of booth seat length measured at the backrest of the seating booth.

MEANS OF EGRESS ILLUMINATION

The means of egress, including the exit discharge, must be illuminated at all times in the building space that the egress occupies, except for the following:

- Occupancies in Group U
- Aisle access ways in Group A
- Dwelling and sleeping units in Groups R-1, R-2, and R-3
- Sleeping units of Group I occupancies.

The illumination level cannot be less than one foot-candle or 11 lux at the walking surface level except for auditoriums, theaters, concert or opera halls and similar assembly occupancies. The illumination at the walking surface level is allowed to be reduced during performances to not less than 0.2 foot-candle, provided that the required illumination is automatically restored upon activation of a premises' fire alarm system where such system is provided.

Did You Know?

This code requires that the total width of means of egress in inches must not be less than the total occupant load served by the means of egress multiplied by the factors in tables found in the code and not less than specified elsewhere in this code. Multiple means of egress much be sized such that the loss of any one means of egress must not reduce the available capacity to less than 50 percent of the required capacity. The maximum capacity required from any story of a building shall be maintained to the termination of the means of egress.

!Definitionalert

Exit: That portion of a means of egress system which is separated from other interior spaces of a building or structure by fire-resistance-rated construction and opening protectives as required providing a protected path of egress travel between the exit access and the exit discharge. Exits include exterior exit doors at ground level, exit enclosures, exit passageways, exterior exit stairs, exterior exit ramps, and horizontal exits.

The power supply for exit illumination must normally be provided by the electrical supply. In the event that the power supply is cut off an emergency electrical system must automatically illuminate the following areas:

- Aisles and unenclosed egress stairways in rooms and spaces that require two or more means of exits
- Corridors, exit enclosures, and exit passageways in buildings required to have two or more exits
- Exterior exit components at other than the level of exit discharge until exit discharge is accomplished for buildings required to have two or more exits
- Interior exit discharge elements, as permitted in this chapter, in buildings required to have two or more exits
- Exterior landings for exit discharge doorways in buildings required to have two or more exits.

ACCESSIBLE MEANS OF EGRESS

Accessible spaces are to be provided with no less than one means of accessible means of egress. Three exceptions to accessible spaces exits are:

- Accessible means of egress are not required in alterations to existing buildings.

Did You Know?

The emergency power system must provide power for the duration of not less than 90 minutes and has to consist of storage batteries, unit equipment, or an on-site generator.

- One accessible means of egress is required from an accessible mezzanine level in accordance with this chapter.
- In assembly spaces with sloped floors, one accessible means of egress is required from a space where the common path of travel of the accessible route for access to the wheelchair spaces meets the requirements in this chapter.

Each required accessible means of egress must be continuous to a public way and has to consist of one or more of the following components:

- Accessible routes
- Stairways within vertical exit enclosures
- Exterior exit stairways
- Elevators
- Platform lifts
- Horizontal exits
- Ramps
- Areas of refuge.

!Definitionalert

Accessible means of egress: A continuous and unobstructed way of egress travel from any accessible point in a building or facility to a public way.

These components must comply with the International Building Code book. At least one required accessible means of egress must be an elevator complying with this section, except for buildings that are equipped throughout with an automatic sprinkler system. The elevator is not required on floors provided with a horizontal exit and located at or above the level of exit discharge or if the building provides a ramp that conforms to this chapter.

The use of stairs for an accessible means of exit is common. In order to use stairs in this capacity, the exit stairway has to have a clear width of 48 inches between handrails and must either incorporate an area of refuge within an enlarged floor-level landing or must be accessed from either an area of refuge or a horizontal exit. Please see the following list for exceptions.

- Unenclosed exit stairways as permitted by this chapter are allowed to be considered part of an accessible means of egress.
- The area of refuge is not required at unenclosed exit stairways as permitted by this chapter in buildings or facilities that are equipped throughout with an automatic sprinkler system.
- The clear width of 48 inches between handrails is not required at exit stairways in buildings or facilities equipped throughout with an automatic sprinkler system installed or for exit stairways accessed from a horizontal exit.
- Areas of refuge are not required at exit stairways serving open parking garages.

In order to be considered part of an accessible means of egress, an elevator must comply with the emergency operation and signaling device requirements or Section 2.27 of ASME A17.1, and standby power must be provided. The elevator must be accessed from either an area of refuge complying with this chapter or a horizontal exit; elevators are not required to be accessed from an area of refuge or horizontal exit in open parking garages.

Lifts that are used for wheelchairs do not serve as part of an accessible means of egress, except where allowed as part of a required accessible route. If you plan on installing a platform lift on an accessible means of egress, they cannot be installed in a fully enclosed or fire-rated shaft.

In your building structure you must have an area of refuge, which is required by code. This refuge space must be accessible from the space it serves by an easily reached means of an exit or way out. Keep in mind that this accessible space to the exit has to have a maximum travel distance and cannot be more than the travel distance that is permitted for the type of occupancy that you have built. Also, every required area of refuge must have a direct access to an enclosed stairway or an elevator.

You may use an elevator lobby for an area of refuge as well, and it must comply with the code regarding smokeproof enclosures. In regards to any refuge, stairway, or elevator and lobby they all must comply with the code in this chapter. Don't be fooled into thinking that just by obeying the code to provide refuge areas is enough. There is a size requirement as well.

Each area of refuge must always be large enough to accommodate one wheelchair space of 30 inches by 48 inches for each 200 occupants, based on the occupant load of the area of refuge and areas that are served by these areas. The wheelchair spaces cannot reduce the required means of egress width and access to any of the required wheelchair spaces in an area of refuge can never be obstructed by more than one adjoining wheelchair space.

A separation of each area of refuge by a smoke barrier is mandated. This barrier must comply with the code or a horizontal exit complying with this chapter. Each area of refuge must be designed to minimize the intrusion of smoke, except for those areas which are located within a vertical exit enclosure. Each area of refuge must have a central control point and must be equipped with a two-way communication system.

If the central control point is not constantly attended, the area has to have controlled access to a public telephone. All locations of the central control point must have a stamp of approval by the fire department. Don't forget

CODE UPDATE

Egress doors are normally required to be of the pivoted or side-hinged swinging type.

CODE UPDATE

With some exceptions, stair treads and risers are required to be of a uniform shape and size.

that when you choose a two-way communication system that it must have both audible and visible signals. Make sure that in any areas of refuge instructions on the use of the areas (under emergency conditions) are posted with the communications system. All instructions must have the following items included in them:

- Directions to find other means of egress
- Persons able to use the exit stairway must do so as soon as possible, unless they are assisting others.
- Information on planned availability of assistance in the use of stairs or supervised operation of elevators and how to ask for such assistance
- Directions for the use of the emergency communications system.

Each door that provides access to an area of refuge from an adjacent floor area must be identified by a sign that complies with ICC A117.1; it must clearly state: AREA OF REFUGE and it also has to include the International Symbol of Accessibility. In places that require an illuminated sign, the area of refuge sign must be illuminated.

DOORS, GATES, AND TURNSTILES

The above is just one definition of a door. This section has many code rules regarding doors, gates, and turnstiles, all of which are used as a means of exit. Means of doors must be readily distinguishable from the adjacent construction and finishes such that the doors are easily recognizable as doors.

Mirrors or similar reflecting materials will not be used on means of exit doors. Means of exit doors will not be hidden by curtains, drapes, decorations, or similar materials.

The minimum width allowed of each door opening has to be sufficient for the occupant load and has to provide a clear width of not less than 32 inches. If the doorway has swinging doors the clear opening has to be measured between the face of the door and the stop, with the door open 90 degrees. Where this section requires a minimum clear width of 32 inches and a door opening includes two door leaves without a mullion, which is a vertical piece of wood that divides the opening, one leaf must provide a clear opening width of 32 inches.

The maximum width of a swinging door leaf has to be 48 inches nominal. Exit doors in a Group I-2 occupancy used for the movement of beds must have a clear width not less than 41.5 inches. One more thing: the height of all doors cannot be less than 80 inches. Review the following list of exceptions of width and height of doors:

- The minimum and maximum width does not apply to door openings that are not part of the required means of egress in Group R-2 and R-3 occupancies.
- Door openings to resident sleeping units in Group I-3 occupancies must have a clear width of not less than 28 inches.
- Door openings to storage closets less than 10 square feet in area are not limited by the minimum width.
- Width of door leafs in revolving doors that comply with this chapter are not limited.
- Door openings within a dwelling or sleeping unit cannot be less than 78 inches in height.
- Outside door openings in dwellings and sleeping units, other than the required exit door, are not to be less than 76 inches in height.

!Definitionalert

Door, balanced: A door equipped with double-pivoted hardware so designed as to cause a semi counterbalanced swing action when opening.

- In other than Group R-1 occupancies, the minimum widths apply to inside exit doors within a dwelling or sleeping unit that is not required to be an accessible unit, Type A or Type B unit.
- Door openings required to be accessible within Type B units must have a minimum clear width of 31.75 inches.

For doors that have the required clear width of less than 34 inches above the ground, there must be no projections. Projections into the clear opening width between 34 inches and 80 inches above the floor or ground cannot exceed 4 inches. Did you know that exit doors must be side-hinged swinging? Read on for exceptions to this rule.

- Private garages, office areas, factory and storage areas with an occupant load of 10 or less
- Group I-3 occupancies used as a place of detention
- Critical or intensive care patient rooms within suites of health care facilities
- Doors within or serving a single dwelling unit in Groups R-2 and R-3
- In other than Group H occupancies, revolving doors complying with this chapter
- In other than Group H occupancies, horizontal sliding doors are permitted as a means of egress
- Power-operated doors
- Doors serving a bathroom within an individual sleeping unit in Group R-1.

All of these exceptions must comply with their proper chapter or section of the code book. Doors must swing in the direction of exit travel where serving an oc-

CODE UPDATE

Exit signs are not required in dayrooms, sleeping rooms, or dormitories in occupancies in Group I-3.

cupant load of 50 or more persons or a Group H occupancy. The opening force for inside side-swinging door closers cannot go over a 5-pound force. For other side-swinging, sliding, and folding doors, the door latch must release when subjected to a 15-pound force. Plus, the door must be set in motion when faced with a 30-pound force. The door must swing to a full-open position when subjected to a 15-pound force. All forces must be applied to the side with the latch. If you have installed special doors, such as revolving doors, and security grilles they must comply with the following list:

- Each revolving door has to be capable of collapsing into a book fold position with parallel exit paths providing an aggregate width of 36 inches.
- A revolving door cannot be located within 10 feet of the foot of or top of stairs or escalators.
- A dispersal area must be provided between stairs or escalators and revolving doors.
- The revolutions per minute for a revolving door cannot be more than those shown in the code.
- Each revolving door must have a side-hinged swinging door which complies with this chapter in the same wall and within 10 feet of the revolving door.

A revolving door that is used as a component of a means of exit must comply with this chapter and meet the following conditions:

- Revolving doors cannot be given credit for more than 50 percent of the required exit capacity.
- Each revolving door must be credited with no more than a 50-person capacity.
- Each revolving door must be capable of being collapsed when a force of no more than 130 pounds is applied within 3 inches of the outer edge of a wing.

If a revolving door is being used as other than a component of a means of egress then the collapsing force of a revolving door not used in this capacity must have a means of egress not more than 180 pounds. An exception to this is a collapsing force in excess of 180 pounds is permitted if the collapsing force is reduced to not more than 130 pounds when at least one of the following conditions is met:

- There is a power failure of power is removed to the device holding the door wings in position
- There is an actuation of the automatic sprinkler system where this system is provided
- There is an actuation of a smoke detection system which is installed to provide coverage within the building which is within 75 feet of the revolving doors
- There is an actuation of a manual control switch, in an approved location and clearly defined, which reduces the holding force to below the 130-pound force level.

FLOORS AND LANDINGS

Floors and landings have to be at the same elevation on each side of a door. Landings must be level, except for outside landings which are allowed to have a slope that does not exceed 0.25 unit vertical in 12 unit horizontal or 2-percent slope. See the list below for exceptions:

- Doors serving individual dwelling units in Groups R-2 and R-3 where a door is permitted to open at the top step of an interior flight of stairs provided the door does not swing over the top step, however, screen and storm doors are permitted to swing over stairs or landings
- In Group R-3 occupancies not required to be accessible units, Type A or Type B units, the landing at an outside doorway must not be more than 7.75 inches below the top of the threshold, provided that the door, other than an outside storm or screen door, does not swing over the landing.
- Variations in elevation due to differences in finish materials, but not more than 0.5 inch
- Outside decks, patios, or balconies that are part of Type B dwelling units, have inaccessible surfaces, and are not more than 4 inches below the finished floor level of the adjacent interior space of the living unit.

DOOR HANDLES

Door handles, pulls, latches, and locks must be installed 34 inches minimum and 48 inches maximum above the finished floor. Locks used only for security purposes and not used for normal operation are allowed at any height except for access doors or gates in barrier walls and fences that are used to protect pools, spas and hot tubs which are allowed to have operable parts so the latch release on self-latching devices is at 54 inches maximum above the finished floor or ground, but they cannot be self-locking devices that need a key, electronic opener, or integral combination lock.

There are doors in certain occupancies where locks and latches are permissible in places of detention or restraint and in buildings in occupancy Group A having an occupant load of 300 or less, Groups B, F, M, and S, and in places of religious worship.

The main exterior of these doors are allowed to be equipped with key-operated locking devices from the exit side as long as the locking device can be identified as being locked. There also needs to be a sign, visible to all, adjacent to the door that clearly states: This Door To Remain Unlocked When The Building Is Occupied. There is no exception to this rule!

The sign letters are to be 1 inch high on a contrasting background. Anyplace where the doors are used in pairs, automatic flush bolts are allowed to be used, but they need to be approved and only allowed if the door leaf that has the automatic flush bolts has no doorknob or surface-mounted hardware.

Did You Know?

Doorway landings must have a width of no less than the width of the stairway or the door, whichever is greatest. Doors in a fully open position must not reduce the landing to less than one-half its required width. Landings must have a length measured in the direction of travel of no less than 44 inches. One exception to this code is that the landing length in the direction of travel in Groups R-3 and U and within individual units of Group R-2 does not have to be more than 36 inches.

Doors that are individual dwelling or sleeping units in Group R occupancies with an occupant load of 10 or less are permitted to be equipped with a night latch, dead bolt, or security chain, but only if they are able to be open from the inside without using a key or tool.

STAIRWAYS

Handrails with a circular cross-section must have an outside diameter of at least 1.25 inches, but no greater than 2 inches or must provide equivalent graspability. If the handrail is not circular, it must have a perimeter dimension of at least 4 inches and not greater than 6.25 inches with a maximum cross-section dimension of 2.25 inches. The maximum radius of edges must be 0.01 inch. Generally speaking, handrail-gripping surfaces must be continuous, without interruption by newel posts or other obstructions.

The code states that the width of stairways cannot be less than 44 inches. There are a few exceptions to this as follows:

- Stairways that serve an occupant load of less than 50 must have a width of no less than 36 inches.
- Where an incline platform lift or stairway chairlift is installed on stairways that serve occupancies in Group R-3, or within dwelling units in occupancies Group R-2, a clear passage width of no less than 20 inches must be provided. If the seat and platform can be folded when not in use, the distance can be measured from the folded position.

All stairways must have a minimum headroom clearance of 80 inches. This has to be measured vertically from a line connecting the edge of the nosings. The headroom has to be continuous from above the stairway to the point where the line intersects the landing below, one tread depth beyond the bottom riser. Note, however, that spiral stairways, while complying with this section, are allowed to have a 78-inch headroom clearance.

Stair tread depths have a minimum requirement of 11 inches. Riser heights are measured vertically between the leading edges of adjacent treads and the tread depth is to be measured horizontally between the vertical planes of the primary projection of adjacent treads and at a right angle to the

!Definitionalert

Stairway: One or more flights of stairs (a change in elevation, consisting of one or more risers, as defined by the code), either exterior or interior, with the necessary landing and platforms connecting them, to form a continuous and uninterrupted passage from one level to another.

tread's leading edge. Winders also have a minimum tread depth which is 11 inches that are measured at a right angle to the tread's leading edge at a point 12 inches from the side where the treads are narrower and a minimum tread depth of 10 inches. It's important for these measurements to be absolutely accurate. Any difference can make the stairways too steep; appear too high or even too long.

Alternating tread devices, spiral stairways, and aisle stairs are a few exceptions, and are discussed later in this chapter. However, in Group R-3 occupancies, within dwelling units in Group R-2 occupancies, and in Group U occupancies that are accessory to a Group R-3 occupancy or accessory to individual dwelling units in Group R-2 occupancies the maximum riser height is 7.75 inches. The maximum winder tread depth at the walk line must be 10 inches and the minimum winder tread depth is 6 inches. A nosing no less than 0.75 inch, but not more than 1.25 inches must be provided on stairways with solid risers where the tread depth is less than 11 inches. These measurements are exact and there are no exceptions. The tolerance between the largest and smallest riser height or between the largest and smallest tread depth must not be more than 0.375 inch in any flight of stairs.

The greatest winder tread depth at the 12-inch walk line within any flight of stairs must not exceed the smallest by more than 0.375 inch measured at a right angle to the tread's leading edge. Where the bottom or top riser adjoins a sloping public way, walkway, or driveway having an established grade and serving as a landing, the bottom or top riser is permitted to be reduced along the slope to less than 4 inches in height, with the variation in height of the bottom or top riser not to exceed one unit vertical in 12 units horizontal (8-percent slope) of stairway width.

The nosings or leading edges of treads at such non-uniform height risers must have a distinctive marking stripe, different from any other nosing marking provided on the stair flight. The distinctive marking stripe must be visible in descent of the stair and must have a slip-resistant surface. Any marking strip must be at least 1 inch wide, but no wider than 2 inches.

Stairway landings must be found at the top and the bottom of each stairway, and the width of such landing cannot be less than the width of the stairway. Every landing must have a minimum dimension measured in the direction of travel equal to the width of the stairway. This dimension does not need to be more than 48 inches where the stairway has a straight run. Please see the following items of exceptions:

- Aisle stairs complying with this chapter
- Doors opening onto a landing cannot reduce the landing to less than one-half the required width. When fully open, the door must not project more than 7 inches into a landing.

The stairway walking surface of treads and landings of a stairway cannot be sloped steeper than one unit vertical in 48 units horizontal in any direction. Stairway treads and landing must have a solid surface and finish floor surfaces must be securely attached. The exception is in Groups F, H, and S occupancies, other than areas of parking structures accessible to the public, openings in treads and landings will not be prohibited provided a sphere with a diameter of 1.125 inches cannot pass through the opening.

All outdoor stairways and outdoor approaches to stairways must be designed so that water will not accumulate on walking surfaces. Enclosures under these stairways such as walls and soffits must be protected by one-hour fire-resistance-rated construction or the fire-resistance rating of the stairway enclosure, whichever is greater. Access to the enclosed space cannot be directly from within the stair enclosure, however, spaces under stairways serving and constrained within a single residential dwelling unit in Group R-2 or R-3 are permitted to be protected on the enclosed side with 0.5-inch gypsum board.

Did You Know?

Wood handrails are permitted for all types of construction.

There will be no enclosed usable space under exterior exit stairways unless the space is completely enclosed in one-hour fire-resistance-rated construction. The open space under exterior stairways will not be used for any purpose.

A flight of stairs cannot have a vertical rise greater than 12 feet between floor levels or landings, except for aisle stairs that comply with this code. I said earlier that spiral stairways cannot be used as a component in the means of egress. There are a couple of instances in which they are allowed to be used, and they are as follows:

- Spiral stairways are permitted to be used as a component in the means of egress only within dwelling units or from a space no more than 250 square feet in area and serving no more than five occupants, or from galleries, catwalks, and gridirons in accordance with code.
- When a spiral stairway is allowed, there are provisions that must be followed. A spiral stairway must have a 7.5 inch minimum clear tread depth at a point 12 inches from the narrow edge. The risers must be sufficient to provide headroom of 78 inches minimum, but riser height cannot be more than 9.5 inches and the minimum stairway width must be 26 inches.

Stairways must have handrails on each side and must comply with this chapter. The following list contains the five exceptions for handrails:

- Aisle stairs complying with this code provided with a center handrail do not need additional handrails.
- Stairways within dwelling units, spiral stairways, and aisle stairs serving seating only on one side are permitted to have a handrail on one side only.

- Decks, patios, and walkways that have a single change in elevation where the landing depth on each side of the change of elevation is greater than what is required for a landing do not require handrails.
- In Group R-3 occupancies, a change in elevations consisting of a single riser at an entrance or egress door does not require handrails.
- Changes in room elevations of only one riser within dwelling and sleeping units in Group R-2 and R-3 occupancies do not require handrails.

If your building or structure has four or more stories in height above grade plane, one stairway must extend to the roof surface, unless the roof has a slope steeper than four units vertical in 12 units horizontal (33-percent slope). In buildings without an occupied roof, access to the roof from the top story must be permitted to be by an alternating tread device.

RAMPS

When you hear the word ramp, many different definitions may come to mind. In this section the definition will be as follows:

The provisions of this section apply to ramps used as a component of a means of egress. Please see the list of exceptions below:

- Other than ramps that are part of the accessible routes providing access or ramped aisles within assembly rooms
- Curb ramps must comply with ICC A117.1
- Vehicle ramps in parking garages for pedestrian exit access do not have to comply with this section when they are not an accessible route serving accessible parking spaces, other required accessible elements, or part of an accessible means of egress.

Ramps that are used as part of a means of egress must have a running slope no steeper than one unit straight up and 12 units across. The slope of the other pedestrian ramps cannot be steeper than one unit straight up in eight units horizontal, except for an aisle ramp in occupancies of Group A.

!Definitionalert

Ramp: A walking surface that has a running slope steeper than one unit vertical in 20 units horizontal with a 5-percent slope.

Ramps must have landings at the bottom of each ramp, points of turning, entrance, exits, and at doors. The slope of a ramp's landing cannot be steeper than one unit, vertically, or a 2-percent slope in any direction. Changes in these levels are not permitted.

EXIT SIGNS

All exits and exit access doors have to be marked by an approved exit that can be seen from any direction, so whether you are walking from the south or north of the building, the exit sign is readable from both directions. Where you place your exit sign is very important and the sign must be placed in an area so that no point in a corridor is more than 100 feet, or the viewing distance for the sign, from the nearest visible exit area.

- Exit signs are not required in rooms or areas that require only one exit or exit access.
- Main exterior exit doors or gates that are obviously and clearly identifiable as exits need not have exit signs where approved by a building official.
- Exit signs are not required in occupancies in Group U and individual sleeping units or dwelling units in Group R-1, R-2, or R-3.
- Exit signs are not required in sleeping areas in occupancies in Group I-3.
- In occupancies in Groups A-4 and A-5, exit signs are not required on the seating side of a stadium or theaters or openings into seating areas where exit signs are provided in the concourse that are readily apparent from the stadium or theater within the seating area in an emergency.

Every sign, including directional signs, must have legible letters without decoration. Letters must be at least 6 inches high, and no less than 0.75 inch wide; less than this is against the code. And there is even a rule for spacing between letters which may not be less than 0.375 inch. Larger signs must have letter widths and spacing in proportion to their height.

HANDRAILS

The code for handrail extensions has changed from the previous code. According to the current International Building Code stair risers have a maximum height requirement of 7 inches and a minimum of 4 inches. What good are stairways and ramps without handrails? According to the 2009 International Building Code a handrail is defined as a horizontal or sloping rail intended for grasping by the hand for guidance or support. Handrails must be sufficiently attached and adequate in strength and be no less than 34 inches, but not more than 38 inches.

Handrail height is to be measured from above the stair tread nosing. When used on ramps handrail height is measured from the surface of the ramp slope. Some handrails have a circular cross-section and must have an outside diameter of at least 1.25 inches, but not greater than 2 inches or must be able to be grasped easily. If the handrail is not circular, it must have a perimeter dimension of at least 4 inches, but not greater than 6.25 inches with a maximum cross-section dimension of 2.25 inches. Any edge has to have a minimum radius of 0.01 inch. Gripping surfaces of all handrails is to be continuous, without interruption by newel posts or any other obstruction. The following are exceptions:

- Handrails within dwelling units are permitted to be interrupted by a newel post at a stair landing.
- A volute, turnout, or starting ease is allowed on the lowest tread in houses only.
- Handrail brackets or balusters that are attached to the bottom surface of the handrail that do not hang over the sides of the handrail within 1.5 inches are not considered obstructions.

There must be a clear space between a handrail and a wall or other surface that must be at least 1.5 inches. Keep handrails, walls, or other surfaces ad-

CODE UPDATE

Rooms or spaces having a floor area larger than 1000 square feet, containing a refrigerant evaporator and maintained at a temperature below 68°F. shall have access to not less than two exits or exit access doors.

jacent to the handrail free of any sharp or abrasive elements, as not doing so is a violation of this code. As with handrails, ramps are required to have a minimum space between rails and the wall and must be at least 36 inches. Also please note that projections into the required width of stairways and ramps cannot be more than 4.5 inches at or below the handrail height.

GUARDS

A guard is a building component or a system of building components located at or near the open sides of elevated walking surfaces that minimizes the possibility of a fall from the walking surface to a lower level. Guards must be placed along the sides of stairways, ramps, and landings that are more than 30 inches above the floor or grade below where the glazing does not meet strength requirements. There are places where guards are not required, including:

- On the loading side of docks or piers
- On the side of stages (from the audience side), and raised platforms, including steps leading up to the stage and raised platforms
- At vertical openings in the performance area of stages and platforms
- At elevated walking surfaces attached to stages and platforms providing access to and utilization of special lighting or equipment
- Along vehicle service pits not accessible to the public
- In assemble seating where guards in accordance with this chapter are permitted and provided.

Guards are used as a protective barrier, but the purpose is defeated if the barriers are not of a certain height. According to the code, a guard cannot be less than 42 inches high, and must be measured vertically above the leading edge of the tread, next to the walking surface or beside a seat. However, for those occupancies in Group R-3, and within individual housing units in occupancies in Group R-2, guards whose top rail also serves as a handrail must have a height of no less than 34 and no more than 38 inches, also measured vertically from the leading edge of the stair tread and nose.

The height in assembly seating areas must be in accordance with this chapter. Guards are in play to prevent you, clients, customers, and the people close to you from getting hurt. Therefore, guards must be installed in screen porches and decks where the walking surface is more than 30 inches above the floor or grade below. These decks and screen porches are of the types that are screened in to prevent insect invasion.

Any component of mechanical equipment that requires service, such as appliances, equipment, fans, or roof hatch openings, that are located within 10 feet of a roof edge or open side of a walking surface and the edge or open side is more than 30 inches above the floor or roof require the installation of a guard that prevents the passage of a 21-inch object.

EXIT ACCESS

An exit access is the part of the means of egress or exit system that leads from any occupied portion of a building or structure to an exit. This can be an aisle or a hallway or other means of a walkway. Egress or throughway cannot pass through rooms that are of high hazard occupancy unless the adjoining rooms are accessory to the area and in Groups H, S, or F occupancy when the adjoining or intervening rooms or spaces are the same or an occupancy group that is less hazardous.

An exit cannot pass through kitchens, storage rooms, closets, or other spaces that are used for similar purposes. Keep in mind that an exit egress

CODE UPDATE

The required width of corridors shall be unobstructed.

is not allowed to pass through a room that can be locked or lead through bedrooms or bathrooms. Speaking of dwellings, when more than one tenant lives on any one floor of a building or structure, each tenant space, dwelling, or sleeping unit must be provided with access to the required exits in a clear aisle that does not pass through another tenant unit.

In Group I-2 occupancies, such as hospitals or nursing facilities, rooms or suites must have an exit access door that leads directly to a corridor. The following are exceptions:

- Rooms with exits doors that lead or open directly to the outside at the ground level
- Patient sleeping rooms are permitted to have one intervening room if that room is not used as an exit access for more than eight patients.
- Special nursing suites are allowed to have one intervening room if they allow direct and constant (visual) supervision by nursing personnel.
- Any room that is not located in a suite where the travel distance is not greater than 100 feet is allowed to have exit travel.

Other occupancies other than Groups H-1, H-2, and H-3, the common path of egress or exit travel is not allowed to be more than 75 feet, but Groups H-1, H-2, and H-3 cannot have a common path of egress travel of more than 25 feet. In Group B, F, and S occupancies exit travel cannot be more than 100 feet, and that is only if the building is equipped with an automatic sprinkler system. Say your tenant space, which is considered a Group B, S, or U occupancy, has an occupant load of not more than 30 people, the length of a common path travel cannot be more than 100 feet.

If your building or structure is of Group I-3 occupancy, the path of travel is limited to 100 feet. And finally, the length of a common path travel in Group R-2 occupancy cannot be more than 125 feet provided you have installed an automatic sprinkler system throughout. Always remember that any automatic sprinkler systems must be installed in accordance with this code.

!Definitionalert

Aisle: An exit access component that defines and provides a path of egress travel.

Aisles

Aisles that serve as part of the exit access in the means of egress or exit system must comply with this next section. Aisles must be provided from all occupied parts of the exit access which contain seats, tables, furnishings, displays, and similar fixtures or equipment. The required width of aisles must be unobstructed. Doors, however, when fully opened, and handrails must not reduce width by more than 7 inches. And, doors in any position must not reduce the required width by more than one-half. Other projections such as trim and similar decorative features are allowed to project into the required width of 1.5 from each side.

For Group M occupancies where seating is located at a table or counter and is adjacent to an aisle or aisle accessway, the measurement must be made to a line of at least 19 inches. This must be measured from the edge of the table or counter. In the case of other side boundaries for aisle or aisle accessways, the clear width must be measured to walls, edges of seating, and tread edges, except that handrail projections are permitted.

If tables or counters have fixed seating, meaning the seats are bolted to the floor, the width of the aisle accessway must be measured from the back of the chair. Table and seating accessways are required to have a minimum of 12 inches of width plus 0.5 inch of width for each additional foot. The exception to this is parts of aisle accessways having a length that is not more than 6 feet and used by no more than four people. Balconies used for egress purposes have to conform to the same requirements as corridors as well.

EXIT AND EXIT ACCESS DOORWAYS

Requirements for the number of exits and exit access doorways are dependant upon a number of details in regard to occupancy.

All required exits must be located in such a way that makes their use obvious. Both interior exit stairways and ramps must be enclosed with fire barriers or horizontal assemblies that comply with the code book. The fire-resistance rating of no less than two hours is required when the unit is four stories or more, or no less than one hour when the unit is less than four stories. Be sure to include basements in this calculation, but not mezzanines. Although obvious, an exit enclosure cannot be used for any purpose other than a means of egress. The exceptions for this are numerous and include:

- In all occupancies, other than Group H and I, and stairway is not required to be enclosed when the stairway serves an occupant load of less than 10 and the stairway complies with either Item 1.1 or 1.2. In all cases however, the maximum number of connecting open stories cannot not be more than two.
  - 1.1 The stairway is open to not more than one story above the story at the level of exit discharge; or
  - 1.2 The stairway is open to not more than one story below the story at the level of exit discharge.
- Exits in buildings of Group A-5 where all portions of the means of egress are essentially open to the outside need not be enclosed.
- Stairways that are used and contained in a single residential dwelling or sleeping unit in Group R-1, R-2, or R-3 occupancies are not required to be enclosed.
- Stairways that are not a required means of egress element are not always required to be enclosed.
- Stairways in open parking structures that serve only the parking structure are not required to be enclosed.
- Stairways in Group I-3 occupancies and means of egress stairways are not required to be enclosed.
- In groups other than Group H and Group I occupancies, a maximum of 50 percent of egress stairways serving one adjacent floor are not required to be enclosed, provided at least two means of egress are provided from both floors that are using the open stairways. And, any floors that are interconnected cannot be open to other floors.

- In groups other than Group H and I occupancies, inside stairway exits that provide service for the first and second stories of a building that has an automatic sprinkler system throughout are not required to be enclosed, but only if there are at least two means of exits. These interconnected floors cannot be open to other stories.

Any penetrations into and any openings through an exit enclosure are not allowed except for required exit doors, equipment, and ductwork that is necessary for independent pressurization, sprinkler piping, standpipes, electrical raceway serving the exit enclosure and terminating at a steel box that is not more than 16 square inches. Such penetrations must be protected. There must not be any penetrations or communication openings, whether protected or not, between adjacent exit enclosures. Equipment and ductwork for exit enclosure ventilation as allowed by this chapter must meet one of the following terms:

- All equipment and ductwork has to be located outside of the building and must be directly connected to the exit enclosure by ductwork enclosed in construction as required by shafts.
- Where the equipment and ductwork is located within the exit enclosure, the intake air must be taken directly from the outdoors and the exhaust air must be discharged directly to the outdoors, or the air must be conveyed through ducts enclosed in construction as required by shafts.
- Where equipment and ductwork is located within the building, this equipment and ductwork must be separated from the rest of the building, including other mechanical equipment, with construction as required for shafts.

In each case, openings into the fire-resistance-rated construction must be limited to those that are needed for maintenance and operation and must be protected by opening protectives for shaft enclosures. And all exit enclosure ventilation systems must be independent of other building ventilation systems.

The outside walls of an exit enclosure must comply with the requirements for exterior walls. In cases where non-rated walls or unprotected openings

CODE UPDATE

Exit passageways shall terminate at an exit discharge or a public way.

enclose the outside of the stairway and the walls or openings are exposed by other parts of the building at an angle of less than 180 degrees, the outside building walls within 10 feet horizontally of a non-rated wall or unprotected opening must have a fire-resistance rating of no less than one hour. And as a reminder, openings within inside walls have to be protected by opening protectives that have a fire protection rating of no less than ¾ hour. This type of construction is to extend vertically—from the ground to a point that is 10 feet—above the uppermost landing of the stairway or to the roof line, but make sure that is the lowest level.

FLOOR NUMBER SIGNS

All floor landings in every inside exit enclosure that connects more than three stories must have a sign. These signs must identify the floor level and the direction to the exit of that stairway. It must also include the availability of the roof access from the stairway for the fire department, and is to be located 5 feet above the floor landing in such a position that is visible when the doors are in the open and closed positions.

SMOKEPROOF ENCLOSURES

A smokeproof enclosure or pressurized stairway must exit into a public space or into an exit passageway, yard, or open space. Any of these must have direct access to a public area. An exit passageway must have no other openings and be separated from the remainder of the building by two-hour fire-resistance-rated construction. The following is a list of exceptions:

- Openings in the exit passageway that serves a smokeproof enclosure are allowed where the exit passageway is protected and pressurized in the same manner as the smokeproof enclosure, and openings are protected as required for access from other floors.

- Openings in the exit passageway serving a pressurized stairway are permitted where the exit passageway is protected and pressurized in the same manner as the pressurized stairway.
- A smokeproof enclosure or pressurized stairway must be allowed to pass through areas on the level of discharge or vestibules as allowed by this chapter.

The access to this stairway must be through an entrance hall or an open balcony, unless there is a pressurized stairway.

HORIZONTAL EXITS

A horizontal exit cannot serve as the only exit from a portion of a building. In buildings where two or more exits are required, not more than one-half of the total number of exits can be horizontal exits. However, if your building is of Group I-2 occupancy, horizontal exits are permitted to make up two-thirds of the required exits.

In Group I-3 occupancies, horizontal exits are allowed to make up 100 percent of the exits required. There must at least 6 square feet of accessible space per occupant on each side of the exit for the total number of people in adjoining compartments. Horizontal exits that connect separation of buildings or refuge areas must provide a fire wall or a fire barrier; both complying with the code and having a fire-resistance rating of no less two hours. Such exit separation must extend vertically through all levels of the building unless the floor assemblies have a fire-resistance rating of no less than two hours with all openings having a protective element. The exception being that a fire-resistance rating is not required at horizontal exits be-

!Definitionalert

Horizontal exits: A path of egress travel from one building to an area in another building on approximately the same level, or a path of egress travel through or around a wall or partition to an area on approximately the same level in the same building, which affords safety from fire and smoke from the area of incidence.

tween a building area and an above-grade pedestrian walkway. The walkway has to provide a distance of more than 20 feet and must comply with Special Construction.

Horizontal exits that contain a refuge area must be a space occupied by the same tenant or public area and each refuge area has to be big enough to accommodate the original occupant load plus any extra people from the adjoining compartment. The anticipated occupant load from the adjoining compartment is based on the capacity of the horizontal exit doors that enter the refuge area. The net floor area allowable per occupancies must be as follows for the indicated occupancies:

- Six square feet per occupant for occupancies in Group I-3
- Fifteen square feet per occupant for ambulatory occupancies and 30 square feet per occupant in non-ambulatory occupancies in Group I-2.

EXTERIOR EXIT RAMPS AND STAIRWAYS

If a building or structure includes outside exit ramps and stairways that are used as part of a required means of door or way of exit, pay close attention to this section as it pertains to such ramps and stairways. If your building or structure is of Group I-2 occupancy, you may not use exterior ramps and stairways as an element of a means of exit.

All exterior exit ramps and stairways must have at least one open side and have a minimum of 35 square feet of total open area bordering each floor level and the level of each intermediate landing. Take note that the required open area must be located on less than 42 inches above the bordering floor or landing level.

These types of ramps and stairways have to be separated from the inside of the building. Remember, any opening is limited to those necessary for exit from spaces that are normally occupied. However, there are several exceptions to this listed below:

- Separation from the interior of the building is not required for occupancies, other than those in Group R-1 or R-2, in buildings that are no more than two stories above grade plane where the level of exit discharge is the first story above grade plane.

- Separation from the interior of the building is not required where the exterior ramp or stairway is served by an exterior ramp and/or balcony that connects two remote outside stairways or other approved exits with a perimeter that is no less than 50 percent open. For the opening to be considered open, it must be a minimum of 50 percent of the height of the enclosing wall, with the top of the openings less than 7 feet above the top of the balcony.
- Separation from the inside of the building is not required for an inside ramp or stairway located in a building or structure that is allowed to have unenclosed inside stairways in accordance with the chapter.
- Separation from the inside of the building is not required for exterior ramps or stairways connected to open-ended corridors, provided that these items are met: the building, which includes corridors and ramps and/or stairs, must be thoroughly equipped with an automatic sprinkler system. The open-ended corridors are connected on each end to an outside exit ramp or stairway and comply with this chapter.
- At any location in an open-ended corridor where a change of direction exceeds 45 degrees, a clear opening of no less than 35 square feet or an outside ramp or stairway is provided. If/when clear openings are provided, they must be located so the accumulation of smoke or toxic gases is minimized.

EMERGENCY ESCAPE AND RESCUE

Basements and sleeping rooms that are located below the fourth story above grade plane must have at least one outside emergency escape and rescue opening. If a basement contains one or more bedrooms, emergency egress and openings are required for each bedroom, but are not required in adjoining areas of a basement. Exceptions include the following:

- Occupancies other than Group R-3 equipped throughout with an approved automatic sprinkler system.
- Occupancies other than Group R-3 that have rooms used for sleeping and have a door that leads to a fire-resistance-rated corridor that has access to two remote exits in opposite directions.

!Definitionalert

Emergency escape: The code defines an emergency escape as an operable window, door, or other device that provides for means of escape and access for rescue in the event of an emergency.

- The emergency escape and rescue opening is permitted to open onto a balcony, provided the balcony provides access to an exit and the dwelling or sleeping unit has an exit that does not open to an atrium.
- Basements that have a ceiling height of less than 80 inches or basements without habitable space and having no more than 200 square feet in floor area are not required to have an emergency escape and rescue windows.
- High-rise buildings with exceptions.

All emergency escape and rescue openings must have a minimum net clear opening of 5.7 square feet, except for grade-floor openings, which must be at least 5 square feet, and both requiring a minimum height of 24 inches and width of 20 inches. The maximum height from the floor in all cases is to be no greater than 44 inches measured from the floor. Keep in mind that all the emergency openings do not mean a thing if they do not work right or cannot be opened from the inside. You can place bars, grilles, or grates over the escape, but they must be removable from the inside without the use of any tool or key. If you choose to install any of the above on an emergency escape you must also have smoke alarms installed as well.

In conclusion of this chapter, it is important that you realize that these provisions are in play for a number of reasons, with safety being one of the most important. As with any chapter of this book, any questions can be addressed to your local code office or building inspector for clarity.

CHAPTER

7

Accessibility

This chapter examines the design and construction of buildings to be accessible to physically disabled persons. All buildings and facilities are to be designed and constructed to be accessible with this code and ICC A117.1. I have included definition alerts and tables for added information. Do not assume that the code is the same from previous code books—there have been changes and it is your responsibility to be aware of such changes.

SCOPING REQUIREMENTS

The following list covers places and sites where accessibility is required:

- Employee work areas must be designed and constructed so that individuals with disabilities can approach, enter, and exit the work area.
- Work areas or portions of that are less than 150 square feet in area and are elevated 7 inches or more above the ground or finish floor are exempt from all requirements, provided that the elevation is essential to the function of that space.

!Definitionalert

Accessible: A site, building, facility, or portion thereof that complies with this chapter.

- Walk-in coolers and freezers intended for employee use only are not required to be accessible.
- Occupancies in Group U are exempt from the requirements of this chapter other than the following: in agricultural buildings, access is required to paved work areas and areas open to the general public, private garages or carports that contain required accessible parking.
- In buildings where a day care facility (Groups A-3, E, I-4, and R-3) is part of a dwelling unit, only the portion of the structure used for the day care facility is required to be accessible.
- The operable parts of fuel-dispensing devices must comply with ICC A177.1, Section 308.2.1 or 308.3.1.

The next list contains sites and places where accessibility is not required. This list and the one above are general lists of sites and places only. They are not intended to take the place of accessibility codes required for routes, entrances, and other requirements that will follow this section.

- Detached one-and two-family dwellings and accessory structures, and their associated sites and facilities, are not required to be accessible.
- Construction sites, structures, and equipment directly associated with the actual processes of construction including, but not limited to, scaffolding, bridging, materials, hoists, or construction trailers are not required to be accessible.
- Raised areas used primarily for purposes of security, life safety, or fire safety including, but not limited to, observation galleries, prison guard towers, fire towers, or lifeguard stands are not required to be accessible or to be served by an accessible route.

- Limited access spaces such as non-occupiable spaces accessed only by ladders, catwalks, crawl spaces, freight elevators, or very narrow passageways are not required to be accessible. This includes, but is not limited to, spaces frequented only by personnel for maintenance, repair, or monitoring of equipment such as elevator pits, water or sewage treatment pump rooms.
- Single-occupant structures accessed only by passageways below grade or elevated above grade including, but not limited to, toll booths that are accessed only by underground tunnels are not required to be accessible.
- Buildings of Group R-1 containing not more than five sleeping units for rent or hire that are also occupied as the residence of the proprietor are not required to be accessible.
- In detention and correctional facilities, common use areas that are used only by inmates or detainees and security personnel, and that do not serve holding cells or housing cells required to be accessible, are not required to be accessible or to be served by an accessible route.

ACCESSIBLE ROUTE

Accessible routes within the site must be provided from public transportation stops, accessible parking areas, accessible passenger loading zones, and public streets or sidewalks to the building's accessible entrance being used. Other than buildings or facilities serving Type B units, an accessible route is not required between site arrival points and the building or facility entrance if the only means of access between them is a vehicular way not providing for pedestrian access.

When a building or portion of a building is required to be accessible, an accessible route must be provided to each portion of the building, to accessi-

CODE UPDATE

At least 10 percent, but not less than one, of patient and visitor parking spaces provided to serve hospital outpatient facilities shall be accessible.

Did You Know?

The only time a single accessible route may pass through a kitchen or storage room is in an accessible Type A or Type B dwelling unit.

ble building entrances that connect each accessible pedestrian walkways, and the public way. However, in assembly areas with fixed seating required to be accessible, an accessible route is not required to serve fixed seating where wheelchair spaces or designated aisle seats required on an accessible route are not provided.

When discussing connected areas and walkways this includes employee work areas, which must have common use circulation paths within such work areas that must be accessible routes. See the following list of exceptions for common use circulation paths:

- Any path located within work areas that are less than 300 square feet and have definite installed partitions, counters, casework, or furniture is not required to be an accessible route.
- Any paths that are an integral component of equipment are not required to be accessible routes.
- Work areas that are fully exposed to the weather are not required to be accessible routes.

An accessible route is not required to stories and mezzanines above and below accessible levels that have a total area of not more than 3000 square feet. This exception does not apply to the following:

- Multiple tenant facilities of Group M occupancies containing five or more tenant spaces
- Levels containing offices of health care providers
- Passenger transportation facilities and airports, which are Group A-3 or B.

In Groups A, I, R, and S occupancies, levels that do not contain accessible elements or other spaces as required by this chapter are not required to be served by an accessible route from an accessible level. If your building is an air traffic control tower or a two-story building that has an occupant load of five or less people (as long as it does not contain public areas) you are not required to have an accessible route.

Accessible routes must happen together with or be located in the same area as a general circulation path. Where the circulation path is inside, the accessible path must be inside as well. If you have only one accessible route in a building, it cannot pass through kitchens, storage rooms, restrooms, closets, or similar spaces.

Accessible routes that are from parking garages and serve Type B dwelling units are not required to be located inside. If you have installed security barriers such as security posts and security check points, they are not allowed to obstruct a required accessible route or accessible means of egress. If the security barriers incorporate elements that cannot comply with these requirements, such as certain metal detectors or other similar devices, the accessible route will be allowed to provide adjacent security screening devices.

Any persons with disabilities who have to pass through one of these devices must be able to maintain visual contact with their belongings to the same extent as others do. One example of where this rule might apply is at an airport security checkpoint when baggage is being checked while going through an x-ray machine. People with disabilities must be able to have the same ability as others to watch their belongings pass through.

ACCESSIBLE ENTRANCES

In addition to the provisions in this chapter regarding accessible entrances, at least 60 percent of all public entrances must be accessible. This does not apply to areas that are not required to be accessible and loading and serv-

!Definitionalert

Public entrance: An entrance that is not a service entrance or a restricted entrance.

ice entrances that are not the only entrance to a tenant space. The following list includes public entrances that must be accessible.

- Parking garage entrances—Direct access for pedestrians from parking structures to buildings or facility entrances must be accessible.
- Entrances from tunnels or elevated walkways—If direct access is provided for pedestrians from a tunnel or elevated walkway to a building or facility, at least one entrance from each tunnel or walkway must be accessible.
- Restricted entrances—In situations where restricted entrances are provided to a building or facility, at least one restricted entrance to the building must be accessible.
- Entrances for inmates or detainees—Buildings with entrances used only by inmates or detainees and security personnel at judicial facilities, detention facilities, or correctional facilities, must have at least one entrance that is accessible.
- Service entrances—If a service entrance is the only entrance to a building or a tenant space in a facility, that entrance must be accessible.
- Tenant spaces, dwelling units, and sleeping units—At least one accessible entrance must be provided to each tenant, dwelling unit, and sleeping unit in a facility, except for the following: an accessible entrance is not required for tenant spaces that are not required to be accessible and an accessible entrance is not required for dwelling units and sleeping units that are not required to be accessible units, Type A units or Type B units.

CODE UPDATE

For every six accessible parking spaces provided at least one of them shall be van accessible.

CODE UPDATE

Mechanical access parking garages are required to provide at least one passenger loading zone at vehicle drop-off and vehicle pick-up areas.

PARKING AND PASSENGER LOADING FACILITIES

For buildings and structures that provide parking, accessible parking spaces must be provided. In cases where more than one parking facility is provided on a site, the number of parking spaces required to be accessible must be calculated separately for each parking facility. Note that this section does not apply to parking spaces that are used exclusively for buses, trucks, other delivery vehicles, law enforcement, or motor pools where lots accessed by the public are provided with an accessible passenger loading zone. The next paragraphs provides detailed information regarding the percentages that different groups or facilities must provide for parking.

For Groups R-2 and R-3, 2 percent, but not less than 1 percent, of each type of parking space provided for occupancies are required to have accessible parking. Type A or B dwelling or sleeping units must be accessible. In locations where parking is provided within or beneath a building, accessible parking spaces must also be provided within or beneath the building. Hospital outpatient facilities must provide 10 percent of accessible patient and visitor parking spaces for hospital outpatient facilities. For rehabilitation facilities and outpatient physical therapy facilities, 20 percent, and not less than one, of the portion of patient and visitor parking spaces must be accessible.

A percentage of accessible parking spaces must be dedicated to van-accessible parking. At least one must be a van-accessible parking space. For every accessible parking space, there are provisions regarding the location of these spaces. Accessible parking spaces must be located on the shortest accessible travel route from adjacent parking to an accessible building entrance. For parking facilities that are not provided for any particular building, accessible parking spaces must also be located on the shortest route to an accessible pedestrian entrance. For buildings that have multiple accessible entrances with adjacent parking, accessible parking spaces must be dispersed and located near the accessible entrances. Pay attention to the following exceptions:

- In multilevel parking structures, van-accessible parking spaces are permitted on one level.
- Accessible parking spaces are allowed to be located in different parking facilities if substantially equal or greater accessibility is provided in terms of distance from an accessible entrance or entrances, parking fee, and user convenience.

DWELLING UNITS AND SLEEPING UNITS

In addition to the other requirements of this chapter, occupancies having dwelling or sleeping units must be provided with accessible features in accordance with this section. Dwelling and sleeping units that are required to be accessible units and Type A and B units must comply with the applicable portions of ICC A117.1. Rooms and spaces that are used by the general public or available for use by residents must be accessible. Some examples of these rooms or spaces are:

- Bathrooms
- Kitchens
- Living and dining rooms
- Patios
- Terraces and balconies.

This list of required rooms does not include recreational facilities in accordance with this chapter. The primary entrance of each of these accessible units, not rooms, must have at least one accessible route connecting the building or facility entrances. However, sometimes there are circum-

CODE UPDATE

All courtrooms shall be accessible.

CODE UPDATE

If a service or facility is provided in an area that is not accessible, the same service or facility shall be provided on an accessible level and shall be accessible.

stances beyond your control. Say that either the slope of the finished ground level between the accessible facility and the building is more than one vertical unit in 12 or if physical barriers or legal restrictions prevent the installation of an accessible route. You are allowed, in accordance with the International Building Code, to install a vehicular route (with parking that complies with this chapter) at each public or common used facility or building in place of the accessible route.

The next section, although brief, is in regards to Group I occupancies. There have been a number of changes for this category that you need to be aware of: At least 4 percent, but not less than one, of the dwelling and sleeping units must be accessible units. Group I-1 structures of Type B units that contain four or more dwelling or sleeping units intended to be occupied as a residence, must be accessible, and the number of Type B units is allowed to be reduced in accordance with this chapter. Group I-2 occupancies, nursing homes, must have at least 50 percent (but not less than one) accessible and Type B units. Accessible and Type B units must be provided for Group I-2, which include the following:

- General-purpose hospitals
- Psychiatric facilities
- Detoxification facilities
- Residential care/assisted living facilities.

Of all the dwelling and sleeping units in these facilities, at least 10 percent, again, no less than one, must be accessible units. For structures with four or more dwelling or sleeping units with the intention to be occupied as a residence, every unit must be a Type B occupancy. However, the number of Type B units is permitted to be reduced in accordance with this chapter.

Did You Know?

Team or player seating at a sports activity must provide at least one wheelchair space, except for team or player seating serving bowling lanes.

Group I-2 is classified as hospitals and rehabilitation facilities. These facilities must specialize in treating conditions that affect mobility, or include units which specialize in such treatment, and 100 percent of these dwelling units must be accessible units. The above code provisions pertain to Group I occupancies only.

Apartment houses, monasteries, and convents belong to Group R-2. Both Type A and B must be provided for this group. If this particular group contains more than 20 percent dwelling or sleeping units, at least 2 percent must be Type A, and all units on site will need to be considered when determining the total number of units. Two exceptions to this are: (1) the number of Type A units is permitted to be reduced and (2) existing structures on a site do not have to be included when calculating the total number of units.

In cases where there are four or more dwelling or sleeping units and they are intended to be occupied as a residence, they must be a Type B unit. Again, the number of Type B units is allowed to be reduced in accordance with this chapter. Not all Group R-2 occupancies are apartment houses, monasteries, and convents. Any other occupancy that falls under Group R-2 must provide accessible and Type B units.

The required number of Type A and Type B units does not apply to a site where the required elevation of the lowest floor or the lowest horizontal structural building members of non-elevator buildings are at or above the design flood elevation which results in the following:

- A difference in elevation between the minimum required floor elevation at the primary entrances and vehicular and pedestrian arrival points within 50 feet over 30 inches

- A slope exceeding 10 percent between the minimum required floor elevation at the primary entrances and vehicular and pedestrian arrival points within 50 feet.

SPECIAL OCCUPANCIES

In addition to other requirements of this chapter, the requirements of this section must be applied to special occupancies. Theaters, bleachers, grandstands, stadiums, arenas, and other fixed seating assembly areas must provide accessible wheelchair spaces that comply also with ICC A117.1. If such stadiums, and the like, provide audible public announcements, they must provide equivalent text information regarding events and facilities. For example, places or areas that have electronic signs must have the ability to display text that is being announced over a PA system, with the exception of announcements that cannot be pre-recorded.

One occupancy which has provisions for accessibility is performance areas. An accessible route must directly connect the performance area to the assembly seating area. Self-service storage areas must provide accessible individual self-storage space.

The final provisions for special occupancies that I want to review are judicial facilities. Each courtroom and central holding cell must be accessible. Where separate central holding cells are provided for adult male and juvenile males and adult and juvenile females, one of each type must be accessible. In places where central holding cells are provided, but are not separated by age or gender, at least one accessible cell must be provided. This is the same case for courtrooms.

CODE UPDATE

In dining areas the dining surfaces for the consumption of food or drink are to have at least 5 percent, but not less than one, designed to be accessible.

OTHER FEATURES AND FACILITIES

A very important area that must adhere to the International Building Code is bathrooms and bathing areas. Not only does this code provide for accessibility, but for facilities provided for both genders. All bathrooms and bathing facilities must be made accessible.

Where a floor level is not required to be connected by an accessible route, the only bathroom or bathing room provided cannot be located on the inaccessible floor. You must provide at least one of each type of fixture, element, control, or dispenser in each accessible bathroom/bathing room. There are several exceptions to these provisions, including:

- Bathrooms or bathing facilities accessed only through a private office and not intended for common or public use are allowed to have doors that swing into the clear floor space, provided the door swing can be reversed.
- The height requirements for the water closet in ICC A117.1 are not applicable.
- Grab bars are not required and the requirements for height, knee, and toe clearance do not apply to a lavatory.
- Where multiple single-user toilet rooms or bathing facilities are clustered at a single location, at least 50 percent but not less than one room for each use at each cluster must be accessible.

The provisions for unisex toilet rooms state that unisex toilet rooms must include only one water closet and only one lavatory. A unisex bathing room in accordance with this chapter will be considered a unisex toilet room. A urinal is allowed in addition to the water closet in a unisex toilet room. Unisex bathing rooms can include only one shower or bathtub fixture and only one water closet and lavatory. An accessible route is mandatory and there must be no more than 500 feet from any separate-sex bathroom to a unisex bathroom.

CODE UPDATE

Jury boxes are required to be provided with a wheelchair space.

PLATFORM LIFTS

Platform lifts are allowed to be part of a required accessible route in new construction where indicated in the following list:

- An accessible route to a performing area and speaker platforms in Group A occupancies
- An accessible route to wheelchair spaces required to comply with the wheelchair space dispersion requirements of this chapter
- An accessible route to spaces that is not open to the general public with an occupant load of not more than five
- An accessible route within a dwelling or sleeping unit
- An accessible route to wheelchair seating spaces located in outdoor dining terraces in Group A-5 occupancies where the means of egress from the dining terraces to a public way are open to the outdoors
- An accessible route to the following: jury boxes, witness stands, raised courtroom stations, judges' benches, stations for clerks, bailiffs, deputies, and court reporters
- An accessible route to load and unload areas serving amusement rides
- An accessible route to play components or safe contained play structures
- An accessible route to team or player seating areas serving areas of sport activity
- An accessible route where existing exterior site constraints make use of a ramp or elevator infeasible.

SIGNAGE

This section focuses on the required accessible elements that are also required by the International Symbol of Accessibility. There are several locations where signs are mandated, including:

- Accessible parking spaces required by this chapter except where the total number of parking spaces provided is four or less

- Accessible passenger loading zones
- Accessible areas of refuge
- Accessible rooms where multiple singe-user toilet or bathing rooms are clustered at a single location
- Accessible entrances where not all entrances are accessible
- Accessible check-out aisles where not all aisles are accessible, the sign, where provided, must be above the check-out aisle
- Unisex bath and bathing rooms
- Accessible dressing, fitting, and locker rooms where not all such rooms are accessible.

Other signs indicating special accessibility provisions are signs that are at each door to an egress stairway, exit passageway, and at all areas of refuge.

CHAPTER

8

Exterior Walls

In this chapter you will find discussions regarding the provisions as set forth by the International Building Code regarding the following items:

- Exterior walls
- Exterior wall coverings
- Exterior wall openings
- Exterior windows and doors
- Architectural trim
- Balconies and similar projections
- Bay and oriel windows.

PERFORMANCE REQUIREMENTS

This section applies to exterior walls, wall coverings, and components of such. All exterior walls need to be weather protected with a weather-resistant exterior wall envelope.

!Definitionalert

Exterior wall envelope: A system or assembly of exterior wall components, including exterior wall finish materials, that provides protection of the building structural members, including framing and sheathing material and conditioned interior space, from the detrimental effects of the exterior environment.

An exterior wall envelope must include flashing and must be designed and constructed in such a manner so that an accumulation of water within the wall assembly is prevented. The envelope must provide a water-resistive barrier behind the exterior veneer, and a way for any water that enters the assembly to the exterior to drain. All protection against condensation in the exterior wall assembly must be provided in accordance with the International Energy Conservation Code.

There are exceptions to weather protection: a weather-resistant exterior wall envelope is not required over concrete or masonry walls that are designed in accordance with code, and compliance with the requirements for a means of drainage are not required for an exterior wall envelope that has been tested and proven to resist wind-driven rain. This includes joints, penetrations, and intersections with different materials, in accordance with ASTM E 331, but only under the following conditions:

- Exterior wall envelope test assemblies must include at least one opening, one control joint, one wall/eave interface, and one wall sill.
- All tested openings and penetrations must be representative of the intended end-use configuration.
- Exterior wall envelope assemblies must be tested at a minimum differential pressure of 6.2 pounds per square foot.
- Exterior wall envelope assemblies must be subjected to a minimum test exposure duration of two hours.

Did You Know?

Exterior wall design has to be considered to resist wind-driven rain and the results of such testing have to indicate that no water could penetrate any control joints in the wall, joints at the perimeter of openings, or intersections of terminations with different materials. The structural aspects of exterior walls and associated openings must be designed in accordance with Structural Design.

If a building or structure is in an area considered to be a flood zone, it is required that exterior walls extend below the design floor elevation and are resistant to water damage. Wood must be treated by a pressure-preservative in accordance with AWPA U1 for the species, product, and end use using a preservative that is found in Section 4 of AWPA U1 or decay-resistant heartwood of redwood, black locust, or cedar.

MATERIALS

All materials used must be of a code-approved type. When constructing a water-resistive barrier a minimum of one layer of asphalt felt, which must be No. 15, complying with ASTM D 226 for Type 1 felt, or other approved materials must be attached to the studs or sheathing. This must include flashing. This is to provide a continuous water-resistive barrier behind the exterior wall veneer.

Did You Know?

In flood hazard areas that are subject to high-velocity wave action, electrical, mechanical, and plumbing system components cannot be mounted on or penetrate through exterior walls that are designed to break away under flood loads.

Did You Know?

If you have chosen to construct exterior walls with cold-rolled copper, the copper needs to conform to the requirements of ASTM B 370 and all lead-coated copper must conform to the requirements of ASTM B 101.

Please refer to the following list for materials and the chapter for which they must be in accordance with:

- Concrete and glass-unit masonry: Exterior walls of concrete construction must be designed and constructed in accordance with code requirements.
- Plastics: Plastic panel, apron, or spandrel walls as defined by this code are not limited in thickness, provided that plastic and their assemblies conform to code requirements and are constructed of approved water-resistant materials of adequate strength to resist the wind loads for cladding.
- Vinyl siding must be certified and labeled as conforming to the requirements of ASTM D 3679 by an approved quality control agency.
- Fiber cement siding must conform to the requirements of ASTM C 1186 and be identified on the label listing an approved quality control agency.

FLASHING

Flashing must be installed in such a manner so as to prevent moisture from entering the wall or to redirect it to the outside. The following list contains all of the perimeters in which flashing must be installed:

- Perimeters of exterior door and window assemblies
- Penetrations and terminations of exterior wall assemblies

!Definitionalert

Flashing: Pieces of sheet metal, or the like, used to reinforce and weatherproof the joints and angles of a roof, as where a roof comes in contact with a wall.

- Exterior wall intersections with roofs, chimneys, porches, decks, balconies, and similar projections
- Built into gutters and similar locations.

Flashing with projection must be installed on both sides and the ends of copings, under sills, and continuously above projecting trim. Moisture is also known to build up in exterior wall pockets and even crevices of buildings and structures. It is crucial that these wall pockets be avoided or protected by using caps or drips.

Flashing and weepholes must be located in the first course of masonry above the finished ground level. This includes the level above the foundation wall or slab. There are other points of support, including structural floors, shelf angles, and lintels where anchored veneers are designed in accordance with this section.

VENEERS

Wood veneers that are found on exterior walls of buildings of Type I, II, III, and IV construction cannot be less than 1 inch thick. Exterior hardboard siding cannot be less than 0.438 inch and must conform the following:

CODE UPDATE

Flashing and weep holes in anchored veneer shall be located in the first course of masonry above finished ground level above the foundation wall or slab, and other points of support.

- The veneer cannot be more than three stories in height, measured from the grade plane.
- Where fire-retardant-treated wood is used, the height cannot be more than four stories.
- The veneer is attached to or furred from a noncombustible backing that is fire-resistance rated as required by other provisions of this code.
- Where open or spaced wood veneers (without concealed spaces) are used, they cannot project more than 24 inches from the building wall.

When using anchored masonry veneer it is important to remember that not only does it have to comply with this code, but also to ACI 530/ASCE 5 and TMS 402. Anchored masonry veneers in accordance with this chapter are not required to meet the tolerances in Article 3.3 G1 or ACI 530.1/ASCE 6/TMS 602, but slab-type veneer units no more than 2 inches in thickness must be anchored directly to masonry, concrete, or stud construction.

If veneer units are of marble, granite, or other stone units of slab and are from ties of corrosion-resistant dowels in drilled holes, they must be located in the middle third of the edge of the units. The spacing of these units cannot be more than 24 inches maximum. Each unit cannot have less than four ties per veneer unit and cannot be more than 20 square feet in area.

If you choose to use veneer ties made of metal, make sure they are smaller in area than 0.0336 by 1 inch or, if made of wire, not smaller in diameter than 0.1483 inch wire. Some people prefer anchored terra cotta or ceramic units to use on exterior walls of buildings. These too have to be installed according to code restrictions.

CODE UPDATE

Where the combustible exterior wall covering is furred from the wall and forms a solid surface, the distance between the back of the covering and the wall shall not exceed 1⅝ inch.

CODE UPDATE

Wood veneer must not exceed 40 feet in height above grade.

Whether terra cotta is tied or not, the minimum thickness is 1.625 inches and it must be directly anchored to masonry, concrete, or stud construction. Tied terra cotta or ceramic veneer units, along with the minimum thickness of 1.625 inches, must also have projecting dovetail webs on the back surface spaced approximately 8 inches. Veneer ties must have sufficient strength to support the full weight of the veneer in tension. Adhered masonry veneer must comply with the applicable requirements in Section 1405.9.1 and Sections 6.1 and 6.3 of ACI 530/ASCE 5/TMS 402.

Metal supports for exterior metal veneer must be protected. You can paint, galvanize, or use another coating or approved treatment. Wood studs, furring strips, or other wood supports for exterior metal veneer must be approved and pressure treated. Joints and edges that are exposed to the weather must be caulked with a durable waterproofing material to prevent moisture from seeping in. Grounding of metal veneers on buildings must comply with the requirements of this code or the ICC Electrical Code.

When using this next type of veneer—glass—there are a number of guidelines you must follow. The area of a single section of thin exterior structural glass veneer cannot be more than 10 square feet in cases where it is not more than 15 feet above the level of the sidewalk or grade level below, and cannot be more than 6 square feet where it is more than 15 feet above that level.

Neither the length nor the height of the glass can be more than 48 inches, nor can the glass be any thicker than 0.344 inch. Only apply the glass veneer after making sure that the backing is thoroughly dry. You must apply an approved bond coat as well. Make sure this is applied uniformly over the entire surface of the backing so as to effectively seal the surface. The glass can then be set in place with mastic cement so that at least 50 percent of the area of each glass unit is directly bonded to the backing by mastic not less than 0.25 inch thick and not more than 0.625 inch thick. The mastic cement must be of an approved kind, and make note that the bond coat and mastic must be evaluated for compatibility and must bond firmly together.

CODE UPDATE

Where open or spaced wood veneers that don't have concealed spaces are used they shall not project more than 2 feet from the building wall.

In areas where glass extends to a sidewalk surface, each section must rest in an approved metal molding and be set at least 0.25 inch above the highest point of the sidewalk. The space between the molding and the sidewalk must be thoroughly caulked and made water tight. If set above sidewalk level and/or above 36 inches, the mastic cement binding must be supplemented with approved nonferrous metal shelf angles located in the horizontal joints in every course. These shelf angles cannot be less than 0.0478-inch thick and no less than 2 inches long and must be spaced at approved intervals; with no less than two angles for each glass unit.

Securing angles to the wall or backing should be done with expansion bolts, toggle bolts, or by other approved methods. If you decide you want to use glass veneers realize that there are specifications for the use of joints. Unless otherwise specifically approved by the building official, abutting areas must be ground square and mitered joints cannot be used except where specifically approved for wide angles.

All glass veneers must be held in place by the use of fastenings at each vertical or horizontal edge, or at the four corners of each glass unit, securing the fastenings to the wall or backing with expansion bolts, toggle bolts, or by other methods. The exposed edges of structural glass veneer must be

Did You Know?

Glazing between the floor and a height of 24 inches must be fixed or have openings that will not allow a 4-inch diameter sphere to pass through. One exception to this is openings that are provided with no window guards that comply with ASTM F 2006 or F 2090.

!Definitionalert

Vinyl siding: A shaped material, made principally from rigid polyvinyl chloride (PVC) that is used as an exterior wall covering.

flashed with overlapping corrosion-resistant metal flashing and caulked with a waterproof compound to prevent moisture from coming between the glass veneer and the backing.

WINDOW SILLS

Where the opening of the sill portion of an operable window is located more than 72 inches above the finished grade, the lowest part of the clear opening of the window must be a minimum of 24 inches above the finished floor surface of the room in which the window is located. This applies to the following list of occupancies:

- Groups R-2
- Groups R-3
- One family
- Two family
- Multiple family dwellings.

VINYL SIDING

Vinyl siding conforming to the requirements of this section and complying with ASTM D 3679 must be allowed to be installed on exterior walls of buildings of Type V construction located in areas where the basic wind speed does not exceed 100 miles per hour and the building height is less than or equal to 40 feet in Exposure C.

In areas where construction is located where the basic wind speed is more than 100 miles per hour, or in building heights are over 40 feet, tests or calculations in compliance with the code must be submitted.

Secure all vinyl siding to the building to protect exterior walls of the building from weather. Siding is to be applied over sheathing or other acceptable materials, and applied to conform to water-resistive barrier requirements in this chapter. Even the nails that you use to apply the siding are specified as to size. Unless otherwise specified in the manufacturer's instruction, nails and accessories must have a minimum of 0.313-inch head diameter and 0.125-inch shank diameter.

All nails must be corrosion resistant and most importantly long enough to penetrate the studs or nailing strip of at least 0.75 inches. If you have chosen fiber cement siding for your building or structure, make sure the panels are installed with the long dimension parallel to framing. Vertical joints must occur over framing members and must be sealed with caulking or covered with battens. Use Z-flashing and solid wood framing for horizontal joints. Lap siding must be lapped a minimum of 1 ¼ inches and the ends must be sealed with caulking covered with an H-section joint cover.

CHAPTER

9

Roof Assemblies and Rooftop Structures

This section covers weather protection. Although short, the importance of becoming familiar with the provisions of this section is not to be taken lightly. I am going to review the provisions of weather protection regarding roof decks first and all others will fall into place from there. I do want to note that in some chapters of this book, I mentioned that some words could be used interchangeably. This chapter, however, contains words and phrases that cannot be used in this manner. For instance, there is a difference in definition between roof coverings, roof deck, and rooftop structure. I have added these as definition alerts for you.

Roof decks are required to be covered with an approved roof covering. These must be secured to the building or structure in accordance with this chapter. All roof coverings must be designed, installed, and maintained in accordance to this code. They must also follow the manufacturer's instructions in that the roof covering will protect the building or structure.

!Definitionalert

Roof deck: The flat or sloped surface not including its supporting members or vertical supports.

Flashing is an integral factor in a building's ability to resist water intrusion and must be installed in such a manner to prevent moisture from entering the wall and roof through joints, materials, and at intersections or through the roof plane. Attention to this important aspect of building performance is necessary in both the design and installation of your building project. Metal flashing must be corrosion resistant with a thickness of no less than 0.019 inch. Please see the following list regarding the locations in which flashing must be installed:

- Wall and roof intersections
- Gutters
- Changes in roof slope or directions
- Around roof openings.

All parapet walls must be properly coped with noncombustible, weatherproof materials. The width of the materials cannot be any less than the thickness of the parapet wall. I'm sure you are aware that a new building or structure will need a roof drainage system. If you didn't know that, you do now. Any design and installation of roof drainage systems must be done in compliance with The International Plumbing Code.

Gutters and leaders that are placed on the outside of buildings have to be made of noncombustible material or a minimum of Schedule 40 plastic

CODE UPDATE

An enclosed, unoccupied structure above the roof of a building, other than a tank, tower, spire, dome cupola, or bulkhead is a penthouse.

pipes. Buildings in Group R-3, private garages, and buildings of Type I construction are not to be included as buildings that must follow is code provision. And all roofs decks must have intake and exhaust vents that are installed according to manufacturer's guidelines.

PERFORMANCE REQUIREMENTS

Wind resistance roof requirements are a serious matter. Just watching the news and hearing about the tremendous damage caused by high winds, heavy rains, and hurricanes is enough for anyone to realize how important the provisions of this section are.

Roof decks and roof coverings must be designed for heavy wind loads. This section will focus more on the testing of roof elements and roof systems and how they must conform to this code.

Roof systems with built-up modified bitumen, either fully adhered or mechanically attached through fastened metal panel roof systems, and other types of membrane roof coverings must be tested in accordance with FM 4450, FM 4470, UL 580, or UL 1897. If you have a metal panel roof system through either fastened or standing seam, also must be tested with UL 580 or ASTM E 1592. An exception to this are metal roofs that are constructed of cold-formed steel, where the roof deck acts as the roof covering and provides both weather protection and support for structural loads. These are permitted to be designed and tested in accordance with the referenced structural design standard of this code. For your information, I have listed the acronyms used in the paragraph above:

!Definitionalert

Roof assembly: A system designed to provide weather protection and resistance to design loads. The system consists of a roof covering and roof deck or a single component serving as both the roof covering and the roof deck. A roof assembly includes the roof deck, vapor retardant, substrate or thermal barrier, insulation, vapor retardant, and roof covering.

Did You Know?

The height of a building is also taken into consideration for the allowance of gravel and stone. The regulations regarding the maximum allowance for roof height permitted for buildings with gravel or stone on the roof in areas outside a hurricane-prone region are addressed in the code.

- FM: Factory Mutual Research
- ASTM: American Society for Testing and Materials.

With these acronyms handy, it will be easier for you to refer to these standards to be sure that your building's components are correctly installed.

The physical properties required for roof coverings installed on low-slope roofs must demonstrate physical integrity over the working life of the roof based upon 2000 hours of exposure to accelerated weather testing that must be conducted in accordance with ASTM G 152, 155, or 154. Roof coverings that are subject to cyclical flexural response due to wind loads cannot demonstrate any major loss of tensile strength.

FIRE CLASSIFICATION

Roof assemblies are divided into classes of A, B, and C. The minimum roof coverings installed on buildings must comply based on the type of construction of the building.

Class A roof assemblies are those that are most effective against severe fire exposure. All Class A roof assemblies and coverings must be listed and identified as such, by an approved testing agency. You may use this class for all buildings or structures of all types of construction, except for Class A roof assemblies with the following coverings:

- Brick
- Masonry
- Slate

- Clay
- Concrete roof tile
- Exposed concrete roof deck
- Ferrous or copper shingles or sheets.

Class B roof assemblies are those that are effective against moderate fire-test exposure and must be listed and identified as such by an approved testing agency. An exception is Class B roof assemblies including those with coverings of metal sheets and shingles.

Class C roof assemblies are those that are effective against light fire-test exposure. Class C assemblies must also be listed and identified by an approved listed agency.

MATERIALS

There are applicable standards that materials must conform to, but if the materials are of a questionable suitability, testing by an approved agency is required by the building official to determine the quality or any limitations of application of the materials.

If you are receiving your roofing materials in bulk shipments there are a couple of things to be aware of. First, all materials must have all information issued in the form of a certificate or on a bill by the manufacturer. Second, all materials must be in packages that have the manufacturer's identifying marks and approved testing agency labels. This is a requirement of this chapter.

REQUIREMENTS FOR ROOF COVERINGS

Provisions of requirements for roof coverings include the way in which roof coverings are applied. This includes the manufacturer's installation

CODE UPDATE

A roof assembly is a system designed to provide weather protection and resistance to design loads.

instructions. For roofs located in areas where the basic wind speed is 110 mph or greater, asphalt shingles must be tested in accordance with ASTM D 3161, Class F.

Asphalt shingles must have self-seal strips or must be interlocking. All packaging must have a label indicating compliance or a listing by an approved testing agency. Now don't think that you can use any type of fasteners for this type of shingle. All fasteners for asphalt shingles must be galvanized, stainless steel, aluminum, or copper roofing nails. These nails must have a minimum 12 gage shank with a minimum diameter head of 0.375 inch.

If roof sheathing is less than 0.75 inch, you must be sure that the nail will penetrate. The code requires that there are a minimum number of fasteners per shingle—no less than four fasteners per strip shingle or two fasteners per individual shingle. There has been a change in the code regarding the underlayment application for roof slopes. Distortions in the underlayment must not interfere with the ability of the shingle to seal.

Roof slopes of four units vertical in 12 units horizontal or greater, underlayment must be one layer applied in the following manner. Underlayment must be applied shingle fashion, parallel to and starting from the eave and lapped 2 inches, fastened sufficiently to hold in place.

As you are aware, various exterior building elements such as roofs, walls, windows, and doors form a packet that protects the inside of the building from elements; water being the biggest threat. Obviously a roof with appropriate drainage minimizes this, but buildings need more than just a

Did You Know?

Distortions in the underlayment must not interfere with the ability of the shingles to seal. In areas of high winds (over 110 mph), underlayment must be applied with corrosion-resistant fasteners, and must be applied along the overlap at a maximum spacing of 36 inches on center, making sure to follow the manufacturer's instructions.

!Definitionalert

Underlayment: One or more layers of felt, sheathing paper, nonbituminous saturated felt, or other approved material over which a steep-slope roof covering is applied.

roof, which brings me to the installation of flashing and why it is so important. Installed at intersecting roofs and parapets, this corrosion-resistant metal overlaps to discourage water entrapment.

Base flashing is the portion of the installation that is attached to the roof itself and cap flashing is attached to the projection or wall. Both must be made of a corrosion-resistant metal.

Valleys occur where the different roof slopes intersect and are considered to be problem spots. Valley linings are to be installed in accordance with the manufacturer's instructions before applying shingles. For open valleys lined with metal, the lining must be at least 16 inches wide. Lining of two plies of mineral-surfaced roll roofing is permitted with the bottom layer having a maximum of 18 inches and the top layer a minimum of 36 inches.

For closed valleys (valleys covered with shingles), lining of just one ply of smooth roll roofing of at least 36 inches wide is permitted.

Be sure to provide a drip edge at eaves and gables of shingle roofs. Overlap to a minimum of 2 inches and extend 0.25 inch below sheathing. Drip edges must be mechanically fastened a maximum of 12 inches.

CODE UPDATE

Roof decks are to be covered with an approved roof covering that is secured to the building or structure it protects.

Did You Know?

Another type of roofing that is used is mineral-surfaced roll roofing that cannot be applied on roof slopes below one unit vertical in 12 units horizontal. You must be careful of the underlayment of this type of roofing.

When you have a tall projection or structure mounted on pitched roofs, such as a chimney, you must use a saddle or cricket. A cricket is actually a ridge that is installed between the roof slope and the structure. Cricket or saddle coverings must be made of sheet metal or of the same material as the roof covering.

In areas where there has been a history of ice formation, an ice barrier that consists of at least two layers of underlayment cemented together or of a self-adhering polymer-modified bitumen sheet must be used instead of normal underlayment and it must extend from the eave's edge to a point at least 24 inches inside the exterior wall line of the building. This does not include detached accessory structures that contain no conditioned floor area. The last type of roof covering is wood shingles.

ROOF INSULATION

The use of above-deck thermal insulation is permitted provided the insulation is covered with an approved roof covering that passes the tests of FM 4450 or UL 1256. This does not include foam plastic roof insulation, which must conform to the material and installation requirements found in the code or where a concrete roof deck is used and the above-deck thermal in-

CODE UPDATE

A cricket or saddle shall be installed on the ridge side of any chimney or penetration greater than 30 inches wide as measured perpendicular to the slope.

!Definitionalert

Penthouse: An enclosed, unoccupied structure above the roof of a building, other than a tank, tower, spire, dome cupola, or bulkhead, occupying not more than one-third of the roof area.

sulation is covered with an approved roof covering. If you are choosing to install cellulosic fiberboard roof installation, you must be sure it conforms to the material and installation requirements.

ROOFTOP STRUCTURES

A penthouse or other projection above the roof in structures of other than Type I construction cannot be more than 28 feet above the roof where used as an enclosure for tanks or for elevators that run to the roof and not more than 18 feet for all other cases. A penthouse, bulkhead, or any other similar projection above the roof cannot be used for any purposes other than housing for mechanical equipment or shelter of vertical shaft openings in the roof.

If a penthouse or bulkhead is being used for purposes other than permitted by this section, the use must conform to the requirements of this code. This section does not prohibit the placing of wood flagpoles on the roof of any building. The type of construction for penthouses must include walls, floors, and roofs, required for that building. Exceptions to this include the following:

- On buildings of Type I and II construction, the exterior walls and roofs of penthouses with a fire separation distance of more than 5 feet and less than 20 feet must be of at least one-hour fire-resistance-rated noncombustible construction.
- Interior framing and walls must be made of noncombustible construction.
- On buildings of Type III, IV, and V construction, the exterior walls of penthouses with a fire separation distance of more than 5 feet and less than 20, must be at least one-hour fire-resistance-rated construction.

CODE UPDATE

Cricket or saddle coverings are to be sheet metal or of the same material as the roof covering.

- Walls with a fire separation distance of 20 feet or greater from a common property line must be of Type IV or noncombustible construction.
- Interior framing and walls must be Type IV or noncombustible construction.
- Unprotected noncombustible enclosures housing only mechanical equipment and located with a minimum fire separation distance of 20 feet will be permitted.
- On one-story buildings, combustible unroofed mechanical equipment screens, fences, or similar enclosures are permitted where located with a fire separation distance of at least 20 feet from adjacent property lines and where not exceeding 4 feet in height above the roof surface.
- Dormers must be of the same type of construction as the roof on which they are placed, or of the exterior walls of the building.

REROOFING

All materials that you use to reroof must be of the same quality as if you were constructing new, except that reroofing does not have to meet the minimum design slope requirements of one-quarter unit vertical in 12 units horizontal for roofs that provide positive roof drainage. Structural roof components must be capable of supporting the roof-covering system and the material and equipment loads of the system. There is a very important provision to remember when reroofing. Never install a new roof without first removing all existing layers of roof coverings where any of the following conditions occur:

- Where the existing roof or roof covering is water soaked or has deteriorated at the point that the existing roof or roof covering is not adequate as a base for additional roofing

Did You Know?

If the application of a new wood shingle or shake roof covering creates a combustible concealed space, you must cover the entire existing surface with gypsum board, mineral fiber, or other approved materials.

- Where the existing roof covering is wood shake, slate, clay, cement, or asbestos-cement tile
- Where the existing roof has two or more applications of any type of roof coverings.

Exceptions to this are:

- Complete and separate roofing systems, such as standing-seam metal roof systems, that are designed to send the roof loads directly to the building's structural system and that does not rely on existing roofs and roof coverings for support, do not require the removal of existing roof coverings.
- Metal panel, metal shingle, and concrete and clay tile roof coverings are allowed over existing wood shake roofs when applied in accordance with this chapter.
- The application of a new protective coating over an existing spray polyurethane foam roofing system is permitted without tear-off of existing roof coverings.

There are some instances where you may reinstall materials. This applies only to slate, clay, or cement tile that is not cracked, damaged, or broken. You must reconstruct flashings in accordance with approved manufacturer's installation instructions as well.

CODE UPDATE

Ice barriers are required on roofs where there is a history of ice build-up on eaves.

CHAPTER

10

Structural Design

As you have seen with past chapters, I don't normally begin with definitions. However there have been changes with these definitions and I have decided the easiest way to point these out is just by giving them their very own section. So, here they are:

- Diaphragm flexible: A diaphragm is flexible for the purpose of distribution of story shear and torsional movement where so indicated in Section 12.3.1 of ASCE 7, as modified in this chapter.
- Fabric partitions: A partition consisting of a finished surface made of fabric, without a continuous rigid backing, that is directly attached to a framing system in which the vertical framing members are spaced greater than 4 feet on center.
- Occupancy category: A category used to determine structural requirements based on occupancy.
- Vehicle barrier system: A system of building components near open sides of a garage floor or ramp or building walls that act as restraints for vehicles.

All of these words have changes in definition, either in whole or part. In this chapter you will notice letters and notations, for example "D + F" may be one that you will see. These notations are also listed as definitions. Please see the following list of notations that you will come across:

- D = Dead load
- E = Combined effect of horizontal and vertical earthquake induced forces as defined in Section 12.4.3 of ASCE 7
- Em = Maximum seismic load effect of horizontal and vertical seismic forces as set forth in Section 12.4.3 of ASCE 7
- F = Loads due to fluids with well-defined pressures and maximum heights
- Fa = Flood load
- H = Load due to lateral earth pressures, ground water pressure, or pressure of bulk materials
- L = Live load, except roof live load, including any permitted live load reduction
- Lr = Roof live load including any permitted live load reduction
- R = Rain load
- S = Snow load
- T = Self-straining force arising from contraction or expansion resulting from temperature change, shrinkage, moisture change, creep in component materials, movement due to differential settlement or combinations thereof
- W = Load due to wind pressure.

CONSTRUCTION DOCUMENTS

Are you aware of what information needs to be included in construction documents? There is some important information that cannot be left out such as the size, section, and relative locations of structural members with floor levels, column centers, and offsets dimensioned. The design loads and other information pertinent to the structural design are described in this section and must be indicated on the construction documents. Please see the following exception: Construction documents for buildings con-

CODE UPDATE

The design load-bearing values of soils shall be shown on the construction documents.

structed in accordance with conventional light-frame construction. Provisions must indicate the following structural design information:

- Floor and roof live loads
- Ground snow lead
- Basic wind speed (three-second gust), miles per hour, and wind exposure
- Seismic design category and site class
- Flood design data, if located in flood hazard areas.

The floor live load that has been uniformly distributed and concentrated and used in the design must be indicated for floor areas. If reduction is allowed, that too, must be indicated. Construction documents must include wind design data, regardless of whether wind loads govern the design of the lateral-force-resisting system of the building.

When wind enters a building, internal pressure and external suction create a combined load on the construction causing the building or structure to result in a pile of sticks. The following list contains commonly used wind data:

- Basic wind speed, miles per hour
- Wind importance factor, and occupancy category
- Wind exposure, if more than one wind exposure it used, the wind exposure and applicable wind direction must be indicated
- The applicable internal pressure coefficient
- Components and cladding.

Earthquake design data is another bit of information that must included. Starting from the seismic importance and occupancy factor to the analysis procedure used, this important information must be included in any con-

Did You Know?

You must indicate any special loads that are applicable to the design of the building. Keep in mind that it is illegal to allow a load greater than permitted to be placed upon any floor or roof of a building. Also, occupancy permits will not be issued until the floor load signs have been installed.

struction documents. If your building is located in whole or part of a flood hazard area, the documentation pertaining to design, if required, must be included and the following information, referenced to the datum on the community's Floor Insurance Rate Map or FIRM, is to be included, regardless of whether flood loads govern the design of the building:

- In flood hazard areas not subject to high-velocity wave action, the elevation of the proposed lowest floor, including the basement; the elevation to which any nonresidential building will be dry flood-proofed
- Elevations to which any nonresidential building will be dry flood proofed
- The proposed elevation of the bottom of the lowest horizontal structural member of the lowest floor, including the basement.

GENERAL DESIGN REQUIREMENTS

Buildings and structures have to be designed and constructed in accordance with the strength design, load and resistance design and allowable stress design as permitted by the applicable material chapters in this book. The structural components of any building or structure must be designed and constructed to safely support the factored loads in load combinations that have been designed in this code. You must be careful not to exceed the appropriated strength limit stated for the materials of construction. You must consult the building official for loads and forces for occupancies or uses that are not covered in the chapter. All structural systems and members must be designed to limit deflections and lateral drift.

Account equilibrium, general stability, and geometric compatibility must be taken into account to determine the load effects on structural members and connections. Residual deformations that tend to accumulate on members under repeated service loads must be included in the analysis in regards to added eccentricities expected to occur during their service life. Any system or method of construction to be used must be based on a rational analysis. This analysis must result in a system that provides a complete load path capable of transferring loads from the origin point to the load-resisting elements.

Any rigid elements that are assumed not to be a part of the lateral-force-resisting system are allowed to be incorporated into buildings provided their effect on the action of the system is considered and provided for in the design. In instances where diaphragms are flexible, or are allowed to be analyzed as flexible, provisions will be made for the increased forces induced on resisting elements of the structural system resulting from torsion due to eccentricity between the center of application of the lateral forces and the center of rigidity of the lateral-force-resisting system.

Buildings and structures that have a relatively low hazard to people if there is an emergency issue are categorized as occupancy I. The categories continue down to IV, with the nature of the occupancy in which these buildings or structures fit. As you can see this doesn't give any information for buildings that have two or more occupancies. In such cases, the structure must be assigned to the classification of the highest nature.

For example, if one part of the building is a storage facility and the other is a jail or detention center, the building must be categorized as occupancy III because there are people who occupy the jail. If the building or structure has two separate portions with separate entrances, they must be categorized separately. The building official is authorized to require an engineer-

CODE UPDATE

Each building and structure shall be assigned an occupancy category.

CODE UPDATE

Handrails and guards shall be designed to resist a load of 50 pounds per linear foot applied in any direction at the top and to transfer this load through the supports to the structure.

ing analysis or a load test, or even both, of any building when there is a question of safety regarding the construction of the occupancy in question.

To resist the uplift and sliding forces that result from the use of prescribed loads you must anchor the roof to walls and columns, and in turn, anchor the walls and columns to foundations. If your building has concrete and masonry walls, these must be anchored to floors, roofs, and other structural elements that are used to provide lateral support for the wall. Realize that the required anchors that you use in masonry walls of hollow units or cavity walls must be embedded in a reinforced grouted structural element of the wall.

Structural elements, components, and cladding must be designed to resist forces from earthquakes and wind with consideration to overturning, sliding, and uplift. As a contractor you must provide continuous load paths for transmitting these forces to the foundation.

LOAD COMBINATIONS

In cases where allowable stress or working stress design is allowed, and used by this code, structures and portions thereof must resist the most critical effects.

Please note these exceptions:

- Crane hook loads do not need to be combined with roof live load or with more than three-fourths of the snow load or one-half of the wind load; and flat roof loads of 30 psf or less do not need to be combined with seismic loads, unless the flat roof snow loads exceed 30 psf, 20 percent must be combined with seismic loads.

- Keep in mind that increases in allowable stresses specified with the referenced standards cannot be used with the load combination listed above, except that a duration of load increase is permitted in accordance with code requirements.
- There may be times when you will be allowed to use alternative basic load combinations that include wind or seismic loads. In these cases, allowable stresses can be increases or load combinations reduced where permitted by the material chapter of this book or the referenced standards.
- Be sure you are confident that you can make these changes and understand the referenced standards carefully.
- Heliports and helistops landing areas are occupancies where the load factors are of a serious nature.

Landing areas must be designed in accordance to the following:

- Dead load, D, plus the gross weight of the helicopter, Dh, plus snow load, S
- Dead load, D, plus two single concentrated impact load, L, approximately 8 feet apart applied anywhere on the landing area, having a magnitude of 0.75 times the gross weight of the helicopter
- Both loads acting together total one and one-half times the gross weight of the helicopter
- Dead load, D, plus a uniform live load, L, of 100 psf.

Please note that there is an exception to this, which is also a change in the code, as follows:

- Landing areas designed for helicopters with gross weights not exceeding 3000 pounds in accordance with bullets one and two (above list) must be allowed to be designed using a 40 psf uniform live load in bullet three (above list), provided the landing area is identified with a 3000 pound weight limitation.
- The 40 psf uniform live load cannot not be reduced and the landing area weight limitation must be indicated by the numeral 3.

!Definitionalert

Dead loads: The weight of materials of construction incorporated into the building, including but not limited to walls, floors, roofs, ceiling, stairways and other similarly incorporated architectural and structural items, and the weight of fixed service equipment, such as cranes, plumbing stacks and risers, electrical feeders, heating, ventilating and air-conditioning systems.

- The appropriate location would be in the bottom right corner of the landing area as viewed from the primary approach.
- The landing area weight limitation must be a minimum of 5 feet in height.

DEAD LOADS

Dead loads are the weights of various structural members and objects that are permanently attached to the structure. There are two types of dead loads—building dead loads and collateral dead loads. A building dead load is the actual building system, such as a roof or floor and materials used for covering such as decking, felt, and hinges. Collateral dead loads are the weight of the permanent materials, such as drywall, sprinklers, and electrical systems. Collateral dead loads do not include the weight of the actual building system. In the absence of definite information, values will be subject to the approval of the building official. We call these dead loads because they are unable to be moved.

LIVE LOADS

One can't have a dead load without a live load, so let's explore this subject a bit more. Live loads are not permanent and can change in magnitude. Live loads include items that can be found inside of a building such as furniture, safes, people, or stored materials. Environmental effects, such as earthquakes, wind, and snow, that have the power to change and cause potential damage or failure to a building, are also considered live loads. However, the live load will be determined in accordance with a method approved by the building official.

Concentrate loads—force localized over a relatively small area such as floors and similar surfaces—must be designed to support live loads that are uniformly distributed. In office buildings and in other buildings where partition locations are subject to change, requirements for partition weight must be made. Whether these partitions are shown on the construction documents or not, you must allow for them unless the specified live load is more than 80 psf. In case you're wondering, the partition load cannot be less than a uniformly distributed live load of 15 psf.

Truck and bus garages are among the many buildings which require a minimum live load. Both the uniform and concentration load must be uniformly distributed over a 10-foot width and placed within their individual lanes. This will produce the maximum stress in each structural member, but not on both at the same time. Keep in mind that all garages accommodating trucks and buses must also be designed with an approved method that contains provisions for traffic railings.

Apartments, which include residential, have requirements for loads. Handrail assemblies and guards must be able to resist 200 pounds in a single concentrated load. This means in any direction at any point along the top, and have attachment devices and supporting structures that can transfer the loading.

With this said and with the exception of roof uniform, live loads and all other minimum uniformly distributed live loads are allowed to be reduced in accordance with certain parts of this section. It is up to you to read and understand these exceptions if they pertain to your building or structure. Note that live loads over 100 psf cannot be reduced unless the following are true:

- The live loads for members supporting two or more floors are permitted to be reduced by a maximum of 20 percent, but the live load cannot be less than the reduced design live load per square foot.
- For uses other than storage, where approved, additional live load reductions are permitted where shown by the registered design professional that a rational approach has been used and that the reductions are necessary.

As an alternative to the section above, floor live loads are permitted to be reduced in accordance with the following provisions. These reductions apply to slab systems, beams, girders, columns, piers, walls, and foundations.

- A reduction is not permitted in Group A occupancies.
- A reduction is not permitted where the live load exceeds 100 psf except that the design live load for members supporting two or more floors is permitted to be reduced by 20 percent.
- A reduction will not be permitted in passenger parking garages except that the live loads for members supporting two or more floors may be reduced by a maximum of 20 percent.
- For all live loads that are not more than 100 psf, the design live load for any structural member supporting 150 square feet or more is allowed to be reduced in accordance with the same code exceptions.

Construction crews must consider not only how loading condition might affect a structure, but also how loads are distributed. This affects both floor and roof loads. Where the uniform floor live load is involved with the design of structural members the minimum applied loads must be the full dead loads on all spans in combination with the floor live loads. This is to create the greatest effect at each location under consideration.

Where uniform roof loads are reduced to less than 20 psf and are involved in the design of structural members, the minimum applied loads must also be the full dead load on adjacent spans or on alternate spans, or whichever produces the greatest effect.

Minimum uniformly distributed roof live loads are allowed to be reduced, but only to the following provision: ordinary flat, pitched, and curved roofs are allowed to be designed for a reduced roof live load. However, you must be careful with this because if your workers are using scaffolding for a work surface for themselves and materials during maintenance and repair jobs, a lower roof load than specified (in the equation below) cannot be used unless approved by the building official. Human safety must always be at the top of your mind.

There are times when cranes are used in construction. The crane live load must always be the rated capacity of the crane. Design loads of moving

!Definitionalert

Wheel load: The vertical force without impact produced on a crane wheel bearing on a runway rail or suspended from a runway beam. Maximum wheel load occurs with the crane at rated capacity and the trolley positioned to provide maximum vertical force at one set of wheels.

bridge cranes and monorail cranes must include the maximum wheel loads of the crane and the vertical impact, lateral, and longitudinal forces made by the moving crane. The design load includes connections and support brackets as well.

Wheel loads of a crane have a maximum load and can be increased by the percentages shown below to determine the induced vertical impact or vibration force.

- Monorail cranes (powered). . .25 percent
- Cab-operated or remotely operated bridge cranes (powered). . .25 percent
- Pendant-operated bridge cranes (powered). . .10 percent
- Bridge cranes or monorail cranes with hand-geared bridge, trolley, and hoist. . .0 percent

SNOW AND WIND LOADS

Other loads include elemental or weather related loads such as snow and wind. Design snow loads are determined in accordance with ASCE 7. You must use the extreme value statistical analysis to determine ground snow loads that are found in the vicinity of the site using a value with a 2-percent annual probability of being exceeded. The ASCE 7 is used to determine wind loads on every building or structure.

The type of opening protection, the basic wind speed, and the exposure category for a site is determined in this chapter. Wind can be assumed to come from any horizontal direction and wind pressures can be assumed to act normal to the surface that is being considered.

There are several areas that are defined as hurricane-prone regions. These regions include the U.S. Atlantic Ocean and Gulf of Mexico coasts where the basic wind speed is greater than 90 mph, and Hawaii, Puerto Rico, Guam, Virgin Islands, and American Samoa. Portions of these areas that are within 1 mile of the coastal mean high water line where the basic wind speed is 110 mph or greater; or portions of hurricane-prone regions where the basic wind speed is 120 mph or greater; or Hawaii are also considered wind-borne debris regions.

For each wind direction considered, an exposure category that adequately reflects The characteristics of ground surface irregularities must be determined for the site at which the building or structure is to be constructed. You must take into account any variation in ground surface roughness that arises from natural topography and vegetation as well as constructed features. A ground surface roughness within each 45-degree sector is determined for a distance upwind of the site as defined from the categories provided below.

- Surface Roughness B: Urban and suburban areas, wooded areas, or other terrain with numerous closely spaced obstructions having the size of single-family dwellings or larger.
- Surface Roughness C: Open terrain with scattered obstructions having heights generally less than 30 feet. This category includes flat open country, grasslands, and all water surfaces in hurricane-prone regions.
- Surface Roughness D: Flat, unobstructed areas and water surfaces outside hurricane-prone regions. This category includes smooth mud flats, salt flats, and unbroken ice.

Roof systems, such as roof decks, are affected by wind loads and must be designed to withstand the wind pressures determined in accordance with ASCE 7. If you are like most people, you've applied or intend to apply asphalt shingles to your roof deck for design. Make sure that any asphalt shingles that you are using haVE been tested to determine the resistance of the sealant to uplift forces using ASTM D 6381. You can find yourself in a

CODE UPDATE

Handrails and guards shall be able to resist a single concentrated load of 200 pounds applied in any direction at any point along the top, and to transfer this load through the supports to the structure.

bit of a mess if you disregard this. All it takes is a heavy wind to come along and blow those shingles off the roof of your building or structure. If you've decided on using rigid tile for your roof covering make sure you note that the code requirement is not the same.

Concrete and clay roof tiles complying with the following limitations must be designed to withstand the aerodynamic uplift moment:

- The roof tiles must be loose laid on battens, mechanically fastened, or mortar/adhesive set.
- The roof tiles must be installed on solid sheathing which has been designed as components and cladding.
- An underlayment must be installed in accordance with code requirements.
- The tile must be single lapped interlocking with a minimum head lap of no less than 2 inches.
- The length of the tile must be between 1.0 and 1.75 feet.
- The maximum thickness of the tail of the tile cannot be more than 1.3 inches.
- Roof tiles using mortar set or adhesive set systems must have at least two-thirds of the tile's area free of mortar or adhesive contact.

SOIL LATERAL LOADS

This section is short, but not necessarily unimportant and refers to basements, foundations, and retaining walls that have to be designed to resist lateral soil loads. Soil loads must be used as the minimum design lateral soil loads unless specified otherwise in a soil investigation report. Make sure you get the investigation report approved by your building official. Basement walls with restricted horizontal movement have to be designed for at-rest pressure. Note that basement walls that do not extend more than 8 feet below grade and supporting flexible floor systems will be allowed to be designed for active pressure.

RAIN LOADS

When designing a building or structure, engineers and architects must consider the loads, both external and internal, that a building must endure. The building is then designed to resist these loads. A type of external load is rain. When designing a building roof the goal is to have each portion of the roof sustain the load of rainwater that will accumulate on it if the primary drainage system is not working properly. This is also true for uniform load caused by water that has risen above the inlet of the secondary drainage system.

A change in the code has a provision for roofs equipped with a slope that is less than ¼ inch per foot. The design calculations must include verification of adequate stiffness to preclude progressive deflection in accordance with Section 8.4 of ASCE 7. One drainage system is simply not enough.

Roofs that are equipped with hardware to control the rate of drainage must be equipped with a secondary drainage system. All secondary drainage systems must be at a higher elevation that limits any accumulation of water on the roof above that elevation.

FLOOD LOADS

Flooding is when excess water overflows from bodies of water onto adjacent land. Flood areas are established by local government bodies with a flood hazard map and other supporting data. This flood hazard map has to include any areas of special flood hazards that have been identified by the Federal Emergency Management Agency and presented as an engineering report. This report has to be written in the following manner: The Flood Insurance Agency for (NAME OF JURISDICTION), dated (INSERT DATE OF ISSUANCE), as amended. Be sure to include the Flood Insurance Rate Map (FIRM) and the Flood Boundary and Flooding Map (FBFM) and any supporting data relating to the flood area.

Floods kill people and destroy homes in parts of the world every year. Obviously, any structures or buildings built in floor-prone areas are permanently at risk. Design and construction of buildings located in flood hazard areas, including those areas that are at risk for high velocity wave action, must be in accordance with ASCE 24. See the following list for documentation that must be prepared and sealed by a registered design professional and submitted to the building official. The first list is for construction in flood hazard areas not to subject high-velocity wave action:

- The elevation of the lowest floor, including the basement, as required by the lowest floor elevation inspection
- For fully enclosed areas below the design flood elevation where provisions to allow for the automatic entry and exit of floodwaters do not meet the minimum requirements in Section 2.6.2.1 of ASCE 24, construction documents must include a statement that the design will provide for equalization of hydrostatic flood forces in accordance with Section 2.6.2.2 of ASCE 24.
- For dry flood-proofed nonresidential buildings, construction documents must include a statement that the dry flood-proofing is designed in accordance with ASCE 24.

For construction in flood hazard areas subject to high-velocity wave action:

- The elevation of the bottom of the lowest horizontal structural member must be as required by the lowest floor elevation inspection.
- Construction documents must include a statement that the building is designed in accordance with ASCE 24, including

CODE UPDATE

When using the alternative all-heights method, wind pressures shall be applied simultaneously on, and in a direction normal to, all building envelope wall and roof surfaces.

that the pile or column foundation and building or structure to be attached is designed to be anchored to resist flotation, collapse and lateral movement due to the effect of wind and flood loads acting simultaneously on all building components, and other load requirements.

- For breakaway walls designed to resist a nominal load of less than 10 psf or more than 20 psf, construction documents must include a statement that the breakaway wall is designed in accordance with ASCE 24.

EARTHQUAKE LOADS

The 2009 International Building Code contains a provision that every structure, and portion thereof, including non-structural components that are permanently attached to structures and their supports and attachments, must be designed and constructed to resist the effects of earthquake motions in accordance with ASCE 7. This does not apply to Exterior Walls, and Appendix 11A of the 2009 International Building Code book. You may use this chapter or ASCE 7 to determine the seismic design category for your structure. Please see the following list of exceptions:

- Detached one- and two-family dwellings that are categorized as A, B, or C of the Seismic Design Category.
- The seismic-force-resisting system of the wood-frame building that conforms to the provisions of the code is not required to be analyzed as specified in this section.
- Agricultural storage structures intended only for incidental human occupancy
- Structures that require special consideration of their response characteristics and environment that are not addressed by this code or ASCE 7 and for which other regulations provide seismic criteria, such as vehicular bridges, electrical transmission towers, hydraulic structures, buried utility lines and their accessories and nuclear reactors.

Any existing buildings that you plan on adding to, making alterations to, or modifying must be done in accordance to Existing Structures. Please note that this applies to any changes in occupancy as well.

!Definitionalert

Site class: A classification assigned to a site based on the types of soils present and their engineering properties as defined in this chapter.

Where required by Special Inspections specifies that seismic requirements include the statement of special inspections and must identify the designated seismic systems and seismic-force-resisting systems and any additional special inspections and testing. This is to include the applicable standards referenced by this code.

Site class definitions are provided in the code. This is the classification assigned to a site based on the types of soils present and their engineering properties. Based on site soil properties, the site is classified as Site Class A, B, C, D, E, or F. But what if you're not sure of the soil properties? If you are unsure of the soil properties of your building or structure site and you do not have sufficient detail to determine such information, use the classification Site Class D until the building official or geotechnical data determines which site class that the soil belongs in.

Consult your local code book for maps of maximum earthquake ground motion for various regions of the world.

CHAPTER

Soils and Foundations

This chapter contains provisions to building and foundation systems in areas that are not subject to scour or water pressure by the actions of wind or wave action.

In this chapter there are allowable bearing pressures, stresses, and design formulas that are used with allowable stress design load combination specifications. When looking at the quality and design of materials used structurally in excavations, footings, and foundations, one must also look at all the requirements that the quality and design materials used for excavations, footings, and foundations must conform to. Excavations and fills must also comply with code requirements.

If the foundation is proportioned using the load combinations allowed, and the computation of the seismic overturning moment is by the equivalent lateral-force method, the proportioning must be in accordance with Section 12.13.4 of ASCE 7.

FOUNDATION AND SOILS INVESTIGATIONS

This section contains provisions for foundation and soil investigations. Classification and investigation of the soil must be made by a registered design professional where required by the building official. If such investigation is required, it is the owner's or applicant's responsibility to submit the foundation and soil investigation to the building official where required in this section.

There are many reasons why a building official may require a foundation and soil investigation. Any soils where the classification, strength, or compressibility are in doubt or where a load-bearing value greater than that specified in this code is claimed, the building official will require the necessary investigation.

There are four provisions that a soil must meet to be considered expansive. Tests that show compliance with the first three bullets will not be required if bullet 4 is conducted.

- Plasticity index (PI) of 15 or greater, determined in accordance with ASTM D 4318
- More than 10 percent of the soil particles pass through a no. 2 sieve, determined in accordance with ASTM D 422
- More than 10 percent of the soil particles are less than 5 micrometers in size, also determined in accordance with ASTM D 422
- Expansion index greater than 20, determined in accordance with ASTM D 4829.

A subsurface soil investigation will be performed to find out if the groundwater table is above or within 5 feet below the elevation of the lowest floor

Did You Know?

Soils that swell when subjected to moisture are classified as expansive soils and contain clay materials that attract and absorb water. If the building official suspects that you may have this type of soil, you can bet that a soil test will be required.

level where floors are located below the finished ground level adjacent to the foundation. A subsurface soil investigation is not required where you can provide waterproofing that is in accordance with this chapter.

Another reason for foundation and soils investigations is for buildings and structures with a Seismic Design Category C, D, E, or F. The investigation for category C differs from D, E, and F and includes an evaluation of the potential hazards that result from earthquake motions such as slope instability, liquefaction, and surface rupture due to faulting or lateral spreading. Categories D, E, and F must meet the same investigation as Category C, in addition to the following:

- To determine the lateral pressures that are made by earthquake motions and the affect they have on basement and retaining walls
- An assessment of potential consequences of any liquefaction and loss of soil strength, including estimation of differential settlement, lateral movement, or reduction in foundation and must address mitigation measures. You may use these measures for consideration in the design of the structure and can include ground stabilization and selection of appropriate foundation types.
- An evaluation for liquefaction and soil strength loss for site peak ground acceleration must be done as well. An exception to this is a site-specific study where peak ground acceleration is determined in accordance with Section 21.2.1 of ASCE 7.
- Foundation and soils investigations include soil classification where soils are grouped according to their general behavior under given physical conditions. It is important that you understand such classifications.

Necessary tests of materials from borings, test pits, or other subsurface exploration and observation are used to determine soil classification. If required, additional studies must be made to evaluate the following:

- Slope stability
- Soil strength
- Position and adequacy of load-bearing soils

- The effect of moisture variation
- Compressibility
- Liquefaction and expansiveness.

The soil classification and design load-bearing capacity must be put on the construction document. The building official may request a written report of the investigation and must be submitted. The following list contains information that is contained in the report. Take note that this list is not limited and there could be more specific information that the building official is requesting.

- A plot showing the location of test borings and/or excavations
- A complete record of the soil samples
- A record of the soil profile
- Elevation of the water table, if encountered
- Recommendations for foundation type and design criteria
- Expected total and differential settlement
- Pole and pier foundation information
- Special design and construction provisions for footings or foundations founded on expansive soils, as necessary
- Compacted fill material properties and testing.

EXCAVATION, GRADING, AND FILL

Before digging any trenches, pits, tunnels, or other excavations, precautions must be taken so that lateral support from any footing or foundation is not disturbed or removed without unpinning or protecting the footing or foundation against settlement or lateral translation.

Backfill is the refilling of an excavated space with soil free of organic material, construction debris, lumps that are larger than a pebble and boulders, or with a controlled low-strength material (CLSM). Backfill must be compacted in a manner that does not damage the foundation or the water/dampproofing materials.

In areas where footings will bear on compacted fill areas you must make sure that the compacted fill complies with the provisions of an approved report. Below is a list of the seven items that must be contained in such approved report:

- Specifications for the preparation of the site prior to placement of the compacted fill material
- Specifications for material to be used as compacted fill
- Test method to be used to determine the maximum dry density and optimum moisture content of the material to be used as compacted fill
- Maximum allowable thickness of each lift of compacted fill material
- Field test methods for determining the in-place dry density of the compacted fill
- Minimum acceptable in-place dry density expressed as a percentage of the maximum dry density determined in accordance with bullet three of this list
- Number or frequency of field tests required to determine compliance with bullet six of this list

The provisions for controlled low-strength material (CLSM) on which footings will bear must follow similar rules. There is an approved report in which specific provisions must be followed, including:

- Specifications for the preparation of the site prior to placement of CLSM
- Specifications for the CLSM

Did You Know?

Compacted fill material that is less than 12 inches in depth does not need to comply with an approved report, provided it has been compacted to a minimum of 90 percent Modified Proctor. See ASTM D 1557 for more information regarding this exception. Also note that controlled low-strength materials do not need to be compacted.

- Laboratory or field test method(s) to be used to determine the compressive strength or bearing capacity of the CLSM
- Test methods for determining the acceptance CLSM in the field
- Number and frequency of field tests required to determine compliance with bullet 4 of this list.

Immediately adjacent to the foundation is where site grading takes place. The slope cannot be any less than one unit vertical for every 20 units horizontal for a minimum distance of 10 feet measured perpendicular to the face of the wall. You must provide a 5-percent slope for an approved alternative if physical obstructions or lot lines prohibit the original 10 feet requirement. When using swales, they must be sloped for a minimum of 2 percent where located within 10 feet of the building foundation.

Resistant surfaces that are within 10 feet of the building foundation must be sloped for a minimum of 2 percent away from the building. One exception to this is for climatic or soil conditions. If such conditions exist, the slope of the ground away from the building foundation is allowed to be reduced, but no less than one unit vertical in 48 units horizontal (2 percent slope). The procedure that you use to establish the final ground level will account for additional settlement of the backfill. Grading and fill for flood hazard areas will not be approved unless the following is true:

- Fill is placed, compacted, and sloped to minimum shifting, slumping, and erosion during the rise and fall of flood water and, if applicable, wave action
- Demonstration through hydrologic and hydraulic analyses performed by a registered design professional and done in accordance with standard engineering practice that the proposed grading or fill will not result in any increased flood levels
- Fill is conducted and/or placed to avoid diversion of water and waves toward any building or structure
- Demonstration that the cumulative effect of the proposed flood hazard area encroachment, when combined with all other existing and anticipated flood hazard area encroachment, will not increase the design flood elevation more than 1 foot at any point.

ALLOWABLE LOAD-BEARING VALUES OF SOILS

There are maximum allowances that you must adhere to for foundation pressure, lateral pressure, or lateral sliding-resistance values. These must not exceed the values allowed by code unless you have data to verify the use of a higher value. Any higher values must be submitted and approved for use.

Do not assume that mud, organic silt, organic clays, peat, or unprepared fill have an acceptable load-bearing capacity unless you have the data to back that up. I believe we all know what happens when we assume something to be true. And it would be a great deal of time, money, and energy wasted if you assume that the use of a material is acceptable without the data to back it up. That being said, there is however, an exception to this. An acceptable load-bearing capacity is permitted to be used if the building official considers the load-bearing capacity of mud, organic silt, or unprepared fill to be adequate for the support of lightweight and temporary structures.

To determine the resistance of structural walls to lateral sliding, calculate by combining the values from the lateral bearing and sliding resistance. Remember you have to submit the reasons or data for this and obtain approval.

In the case of clays, such as sandy, silty, or clayey silt, under no circumstance can the lateral sliding resistance be more than one-half of the dead load. It is possible for increases to be allowed for lateral sliding resistance. For each additional foot of depth to a maximum of 15 times the tabular value.

FOOTINGS AND FOUNDATIONS

Footings and foundations are crucial to the success of a building. When properly designed, footings and foundations withstand the forces of a building. Footings and foundations are built directly on undisturbed soil, compacted fill material, or CLSM, with a minimum depth of footings below the undisturbed surface of 12 inches. While the top surface of footings has to be level, the bottom surface is not. The bottom surface of footings is allowed to have a slope not to exceed one unit vertical in 10 units horizontal.

There are times when it is necessary to change the elevation of the top surface; this is when footings must be stepped. Another important element regarding footings and foundations is frost protection. Foundation walls, piers, and other permanent supports of buildings and structures must be protected, by either extending below the frost line, erecting on solid rock, or constructing in accordance with ASCE 32. If your building is free-standing and all of the following conditions are met, frost protection is not required.

Buildings classified in Occupancy Category I include areas of 600 feet or less for light-frame construction or 400 square feet or less for other than light-frame construction with an eave height of 10 feet or less.

Footings that are on granular soils (soils consisting mainly of sands and gravels), must be located so that the line drawn between the lower edge of adjoining footings will not have a steeper slope more than 30 degrees with the horizontal, unless the material supporting the higher footing is braced or retained or otherwise laterally supported in an approved manner. For the most part, buildings that are below slopes are set apart from the slope at an acceptable distance from the slope for protection from slope drainage.

In cases where the existing slope is steeper than one unit vertical in one unit horizontal, the toe of the slope is to be assumed to be at the foundation.

Most wet basements are caused by surface water which is not adequately drained from the foundation wall. On graded sites, the top of the exterior foundation is to extend above the elevation of the street gutter from an approved drainage device. Alternate elevations are allowed with permission of the building official if you can demonstrate that required drainage to the point of discharge and away from the structure is provided at all locations.

An alternate setback and clearance are also allowed, but, once again, with the building official's permission. The official has the authority to request an investigation and recommendation of a registered design professional to demonstrate that the intent has been satisfied. Any investigation of this request includes the consideration of material, height of the slope and slope gradient, load intensity, and erosion characteristics of any materials used for the slope.

Footings are to have a minimum width of 12 inches. Structural Design, contains provisions for unfavorable effects due to the combinations of load and footings. The dead load is permitted to include the weight of foundations, footings, and overlying fill. Reduced live loads are allowed to be used in the design of footings.

When machinery operations or other vibrations are sent through the foundation, you must give consideration in the footing design to prevent detrimental disturbances of the soil. There is much to learn and understand about vibratory loads to concrete footings. Design, materials, and construction of concrete footings not only have to comply with this chapter, but to concrete requirements as well. Please be aware though, that in instances where a specific design is not provided, concrete footings that support walls that are made of light-frame construction can be designed in accordance with the code, and all concrete in footings must have a specified compressive strength of no less than 2500 pounds per square inch at 28 days.

It is important that you do not place concrete footings through water. You may seek permission from the building official if you are using a funnel or some other method. The key is to get the building official to give you approval to use a funnel to place concrete. You must be sure to protect concrete from freezing during the time that you are placing it and for a total of no less than five days thereafter. Under no circumstance is water allowed to flow through concrete that has been laid in the ground. Next we will examine foundation walls.

When you have a difference in height between the exterior finish ground level and the lower of the top of the concrete footing that supports the foundation you end up with an unbalanced backfill height. There are instances when an unbalanced backfill height is permitted, such as where an interior concrete slab on grade is provided and is in contact with the interior surface of the foundation wall. And foundation walls of rough or random rubble stone cannot be less than 16 inches thick and rubble stone can-

Did You Know?

Plain concrete footings that support walls for something other than light-frame construction cannot have an edge thickness of less than 8 inches placed on soil. The exception being that plain concrete footings that support Group R-3 Occupancies can have an edge thickness of only 6 inches, but only when the footing does not extend beyond the thickness of the footing on either side.

not be used for foundations for structures in Seismic Design Category C, D, E, or F.

Foundation wall materials such as concrete must be constructed in accordance with the code and with the following list:

- The size and spacing of vertical reinforcement is based on the use of reinforcement with minimum yield strength of 60,000 psi, or vertical reinforcement with minimum yield strength of 40,000 psi. 50,000 psi is permitted, provided the same size bar is used and the spacing shown in the table is reduced by multiplying the spacing by 0.67 or 0.83.
- Vertical reinforcement, when required, must be placed nearest the inside face of the wall a distance from the outside face of the wall. The distance is equal to the wall thickness, minus 1.25 inches plus one-half the bar diameter. The reinforcement must be placed within a tolerance of 3/8 inch where the distance is less than or equal to 8 inches.
- Instead of the support shown in some parts of the code, smaller reinforcing bar sizes with closer spaces can be used if this provides an equal amount of reinforcement.
- Any concrete covering used for reinforcement that is measured from the inside face of the wall cannot be less than ½ inch. If measured from the inside face of the wall, the measurement cannot be less than 1.5 inches for no. 5 bars; 2 inches for smaller bars.
- Concrete must have a specified compressive strength of not less than 2500 psi at 28 days.

Just like concrete foundation walls, masonry foundation walls must comply with similar standards:

- The minimum vertical reinforcement for masonry foundations will have a strength of 60,000 psi.
- The specified location of the reinforcement has to be equal or greater to the depth distance as stated in the code.
- Masonry units must be installed with Type M or S mortar.

PIER AND PILE FOUNDATIONS

This section starts with some definitions that pertain to the code requirements of the International Building Code book. You will see these words used frequently in this section so it's good that you have the definitions to refer to when needed.

- Flexural length is the length of the pile from the first point of zero lateral deflection to the underside of the pile cap or grade beam.
- Micropiles are 12-inch-diameter or less bored, grouted-in-place piles incorporating steel pipe (casing) and/or steel reinforcement.
- Pier foundations consist of isolated masonry or cast-in-place concrete structural elements extending into firm materials.
- Piers are relatively short in comparison to their width, with lengths less than or equal to 12 times the least horizontal dimension of the pier. Piers derive their load-carrying capacity through skin friction, end bearing, or a combination of both.
- Pile foundations consist of concrete, wood, or steel structural elements either driven into the ground or cast in place.
- Piles are relatively slender in comparison to their length, with lengths exceeding 12 times the least horizontal dimension.
- Piles derive their load-carrying capacity through skin friction, end bearing or a combination of both.

The general requirements for piers and piles must follow the provisions of this section where Group R-3 and U occupancies do not exceed two stories of light-frame construction or where the surrounding foundation materials furnish adequate lateral support for the pile. These are subject to the approval of the building official. You must design and install your pier and pile foundations on the basis of a foundation investigation (this was defined earlier in the chapter), unless you have sufficient data on which to base your design and installation. An investigation and report is expanded for pier and pile foundations to include the following list (keep in mind though that this list is not limited and the building official may require additional information):

- Recommended pier or pile types and installed capacities
- Recommended center-to-center spacing of piers or piles
- Driving criteria
- Installation procedures
- Field inspection and reporting procedures (to include procedures for verification of the installed bearing capacity where required)
- Pier or pile load test requirements
- Durability of pier or pile materials
- Designation of bearing stratum or strata
- Reductions for group action, where necessary.

There are special types of piles that are not specifically mentioned in this code that can be used. However, you must submit acceptable test data, calculations, and other information that relates to the structural properties and load capacity of such piles. It is only after the building official reviews and approves such information that you will be allowed to use the special pile. In any case, the allowable stresses cannot exceed the limitations that have been set.

You must brace all piers and piles to provide lateral stability in all directions. To be considered braced, three or more piles must be connected by a rigid cap, provided that the piles are located in radial directions from the center of mass of the group not less than 60 degrees apart. A two-pile group in a rigid cap is also considered to be braced along the axis connecting the two piles. Any methods that are used to brace piers or piles must be subject to the approval of the building official. Be careful when installing piers and piles.

Any disturbance in the required sequence of the required installation can cause distortion and damage and can adversely affect the structural integrity of piles that you are currently installing or ones that are already in place. The International Building Code will allow you to reuse existing piers or piles under certain circumstances and only with the approval of the building official. You will have to submit evidence that the piers or piles are sound and meet all requirements of this code.

All exiting piers and piles have to be load tested or redriven to verify their capabilities as well. The design load applied to such piers or piles must be the lowest allowable load as determined by such tests. There is an approved formula along with load tests or methods of analysis to determine the allowable axial and lateral loads on piers or piles. The allowable compressive load on any pile where determined by the application of an approved driving formula can not be more than 40 tons. You must use the wave equation method of analysis for allowable loads over 40 tons to estimate pile drivability of both driving stresses and net displacement per blow at the ultimate load. To use a follower you must obtain permission from the building official and you cannot use a fresh hammer cushion or pile cushion material just prior to final penetration.

When there is any doubt regarding design load for any pier or pile foundation, testing must be done in accordance with ASTM D 1143 or ASTM D 4945. The following are allowable methods of load test evaluations that are permitted to be used:

- Davisson Offset Limit
- Brinch-Hansen 90 Percent Criterion
- Butler-Hoy Criterion
- Other methods approved by the building official.

Piers, individual piles, and group of piles must develop ultimate load capacities of at least twice the design working loads in the designated load-bearing layers. And load-bearing capacities of piers or piles that are discovered to have a sharp or sweeping bend will be determined by an approved method of analysis or by load testing a representative pier or pile. The maximum compressive load on any pier or pile due to mislocation cannot be more than 110 percent of the allowable design load. All piers and piles need proper lateral support to prevent buckling and to allow the design of the pier or pile. This support must be in accordance with accepted engineering practice and provisions of this code.

In addition to learning about allowable pier and pile loads one must be aware of the requirements for seismic design of piers and piles. In the first part of this section you will learn about Seismic Design Category C. (Categories D, E, and F will follow shortly thereafter). Individual pile caps, piers, or piles must be interconnected by ties. These ties must be capable of carrying a force equal to the product of the larger pile cap or column load times the seismic coefficient divided by 10. This can be disregarded only if you can demonstrate that equal restraint is provided with reinforced concrete beams within slabs on grade, reinforced concrete slabs on grade, or very dense granular soils. The code provides an exception to this that states, "Piers supporting foundation walls, isolated interior post detailed so the pier is not subject to lateral loads, lightly loaded exterior decks and patios of Group R-3 and U occupancies not exceeding two stories of light-frame construction, are not subject to interconnection if it can be shown the soils are of adequate stiffness, subject to the approval of the building official."

Seismic category structures must connect concrete piles and concrete-filled steel pipe piles to the pile cap by embedding the pile reinforcement or field-placed dowels anchored in the concrete pile. This must be embedded for a distance that is equal to the development length. For deformed bars this means the development length is the full development length for a compression or tension. You must be sure that the ends of hoops, spirals, and ties are terminated with seismic hooks.

The American Concrete Institute (ACI) 318, Building Code Requirements for Structural Concrete, Section 21.1 defines this with more clarity and I recommend that you refer to this when terminating these ends. Please note that anchorage of concrete-steel pipe piles is allowed to be done using deformed bars developed into the concrete portion of the pile. Structures that are assigned to Seismic Design Category C must follow design details.

Pier or pile moments, shears, and lateral deflections used for design have to be established considering the nonlinear interaction of the shaft and soil, as recommended by a registered design professional. A pile may be

Did You Know?

Unbraced piles in air, water, or in fluid soils have to be designed as columns in accordance with the provisions of this code. Piles that are driven into firm ground are considered to be fixed and laterally supported at 5 feet below the ground surface; in soft material at 10 feet. The building official does have the authority to make any alterations to this after a foundation investigation by an approved agency.

assumed to be rigid if the ratio of the depth of embedment of the pile-to-pile diameter or width is less than or equal to six. You must always include pile group effects from soil on lateral pile nominal strengths where pile center-to-center spacing in the direction of lateral force is less than eight pile diameters. The same is true for vertical pile strength where center-to-center spacing is less than three pile diameters.

The second part of seismic design of piers of piles is for those structures that have be assigned to Seismic Design Category D, E, or F. In addition to this section, Seismic Design Categories D, E, or F must adhere to the requirements for Seismic Design Category C too. Provisions of The American Concrete Institute (ACI) 318, Building Code Requirements for Structural Concrete, Section 21.10.4 must apply when not in conflict with this chapter. Concrete for category D, E, or F must have a specified compressive strength of not less than 3000 psi at 28 days. Please see the following list for exceptions to the above:

- Group R or U occupancies of light-frame construction and two stories or less in height are allowed to use concrete with a specified compressive strength of not less than 2500 psi at 28 days.
- Detached one- and two-family dwellings of light-frame construction and two stories or less in height are not required to comply with the provisions of ACI 318, Section 21.10.4.
- Section 21.10.4 of ACI 318 does not apply to concrete piles.

The design details of piers, piles, and grade beams have to be designed and constructed to withstand maximum imposed curvatures from earthquake and ground motions and structure response. Curvatures must include free-field soil strains that have been modified from soil-pile-structure interaction. Site Class E or F sites have to be designed and detailed in accordance with ACI 318, Sections 21.4.4.1, 21.4.4.2 and 21.4.4.3 within seven pile diameters of the pile cap and the interfaces of soft to prestressed concrete piles. ACI 318 dictates many provisions regarding seismic design, including grade beams. However, grade beams that have the capacity to resist the forces from load combinations do not need to conform to ACI 318.

For piles that are required to resist uplift forces or provide rotational restraint, design of anchorage of piles into the pile cap has to be provided considering the combined effect of axial forces. The minimum of 25 percent of the strength of the pile in tension must include anchorage. Anchorage into the pile cap must be capable of developing the following:

- In the case of uplift, the lesser of the nominal tensile strength of the longitudinal reinforcement in a concrete pile, or the nominal tensile strength of a steel pile, or the pile uplift soil nominal strength factored by 1.3, or the axial tension force resulting from the load combinations
- In the case of rotational restraint, the lesser of the axial and shear forces and moments resulting from the load combinations or development of full axial and shear nominal strength of the pile.

If the vertical lateral-force-resisting elements are columns, the grade beam or pile cap flexural strengths must exceed the column flexural strength. The connections between batter piles and grade beams or pile caps must be designed to resist the nominal strength of the pile acting as a short column. Batter piles and connections must be capable of resisting forces and moments from the load combinations.

DRIVEN PILE FOUNDATIONS

Timber is strong, light in weight, and capable of adequate support. Timber piles are round, tapered timbers with the small end embedded into the soil. Timber piles used to support permanent structures are to be treated in accordance with this section and must be designed in accordance with AFPA NDS (American Forest and Paper Association). Round timber piles must conform to ASTM D 25 while sawn timber piles must conform to DOC PS-20 (Department of Congress). Timber piles that are used for support in permanent structures must comply with this section. If it is established that you will be using the tops of the untreated timber piles below the lowest ground-water level assumed to exist during the life of the lowest structure, only then do timber piles not have to comply with this section.

The AWPA U1 (Commodity Specifications E, Use Category 4C) contains very important information to refer to for driven pile foundations. When working with timber piles and suddenly noticing an increase in rate of penetration you must conduct an investigation for possible damage. If the sudden increase in rate of penetration is not related to soil strata, you must remove the pile for inspection, or if non-viable the timber pile must be rejected.

The second type of driven pile foundations that we will review is precast concrete piles. Precast concrete piles have to comply with design and manufacture, must be of a minimum dimension, and must comply with reinforcement and installation requirements.

To resist all stresses brought on by handling, driving, and service loads; piles must be designed and manufactured in accordance with accepted engineering practices. Concrete piles must have a minimum lateral dimension of 8 inches and corners of square piles must be chamfered, which means to have a bevel or groove and the longitudinal reinforcement must be at least 0.8 percent of the concrete section and consist of at least four bars. You must never drive a precast concrete pile before the concrete has attained a compressive strength of at least 75 percent of the 28-day specified compressive strength, but no less than the strength sufficient enough to withstand handling and driving force.

MICROPILES

Micropiles are 12-inch-diameter or less bored, grouted-in-place piles incorporating steel pipe (casing) and/or steel reinforcement. There has been a change in the code regarding micropiles and in this section I have covered all of these changes. Keep your eyes open for any details that pertain to your construction or building needs and as always ask your local building official to clarify any questions that you may have regarding this code.

Micropiles must have a grouted section reinforced with steel pipe or steel reinforcing. Micropiles develop their load-carrying capacity through soil, bedrock, or a combination of soil and bedrock. The full length of the micropile must contain either a steel pipe or steel reinforcement. One of the materials used with micropiles is grout.

Grout must have a 28-day specified compressive strength no less than 4,000 psi. As with all piles, micropiles too must be reinforced. For piles or portions of piles grouted inside a temporary or permanent casing or inside a hole drilled into bedrock, the steel pipe or reinforcement must be designed to carry at least 40 percent of the design compression load.

You can use rotary or percussive drilling as a method, with or without casing, to form a hole for the pile. The pile must be grouted using a fluid cement grout and pumped through a tremie pipe that extends to the bottom of the pile until the grout comes back up to the top. There are eight requirements of this code that must be applied to specific installation methods:

- For piles grouted inside a temporary casing, the reinforcing steel must be inserted prior to withdrawal of the casing.
- The casing must be withdrawn in a controlled manner with the grout level maintained at the top of the pile to ensure that the grout completely fills the drill hole.
- Make sure you monitor the grout level inside the casing when you are withdrawing the casing so you can see that there is nothing obstructing the flow of the grout.
- You must verify the design diameter of the drill hole for a pile that is grouted in an open drill hole in soil without temporary casing.
- By using a suitable means for piles designed for end bearing you will be verifying that the bearing surface is properly cleaned prior to grouting.
- Subsequent piles cannot be drilled near piles that have been grouted until the grout has had enough time to harden.
- You must grout piles as soon as possible after you have completed drilling.
- For piles designed with casing full length, the casing must be pulled back to the top of the bond zone and reinserted to verify grout coverage outside the casing.

PIER FOUNDATIONS

Isolated piers used as foundation must comply with minimum dimensions of 2 feet with the height not exceeding 12 times the least horizontal dimension. Reinforcements where required must be assembled and tied together and must be placed in the pier hole as a unit before the reinforced portion of the pier is filled with concrete. This does not apply to steel dowels that have been embedded 5 feet or less in the pier. Please note this exception: Reinforcement is permitted to be wet set and the 2 ½ -inch concrete cover

Did You Know?

Where a steel pipe is used for reinforcement, the portion of the cement grout enclosed within the pipe is permitted to be included at the allowable stress of the grout. The provisions for seismic reinforcement differ from above. Any building or structure that is deemed to be of Seismic Design Category C must have a permanent steel casing from the top of the pile down 120 percent times the flexural length. If a building or structure is of Seismic Design Category D, E, or F, this pile will be considered as an alternative system.

requirements can be reduced to 2 inches for Group R-3 and U occupancies that are not more than two stories of light-frame construction, provided that the construction method can be demonstrated to the satisfaction of the building official.

When placing concrete you have to do so in such a way that any foreign matter is taken out and to secure a full-sized shaft. You may not place concrete through water unless a tremie or other method has been approved. Do not just chute the concrete directly into the pier. Concrete must be poured in a rapid and continuous operation through a funnel hopper that you have placed in the center at the top of the pier. If you find that the pier foundation has belled at the bottom, you must check to see that the edge thickness of the bell is not less than what is required for the edge of footings.

This chapter taught us many things about soils and foundations and the do's and don't of piers and pile foundations. It has referred you to other publications which go into greater detail regarding concrete and the provisions that you must follow while building your foundations.

CHAPTER

12

Wood

This chapter contains provisions for materials, design and construction, and the quality of wood and wood fasteners.

There has been a change in the code regarding the general design requirements of structural elements or systems constructed in part or in whole of wood or wood-based products. These must be in accordance with the following:

- Sections 2304, 2305, and 2306 contain the requirements for allowable stress design.
- Sections 2304, 2305, and 2307 contain the requirements for load and resistance factor design.
- Sections 2304 and 2308 contain the requirements for conventional light-frame construction.
- Note this exception: Buildings that have been designed in accordance with the provisions of the AF&PA WFCM are considered to have met the requirements of Section 2308.

Dimensions of lumber are reasoned to be nominal dimensions unless selected as actual dimensions for the purpose of this chapter.

MINIMUM STANDARDS AND QUALITY

This section contains the minimum standards and quality for the different types of wood and other materials used in construction and include, but are not limited to: preservative-treated wood; structural log members; prefabricated wood I-joists; staples; and nails.

Sawn lumber used for load-supporting purposes must be identified by the grade mark of a lumber grading agency. You can use an inspection agency too, but keep in mind that the agency you use must be approved by an accreditation body that complies with DOC PS 20 or equal. All grading practices and identification has to comply with the procedures of DOC PS 20, as well. A certificate of inspection may be accepted instead of the grade mark, but only if all provisions have been met. You are also permitted to use approved end-jointed lumber instead of solid-sawn members as long as they are of the same species and grade.

There are many items in this chapter that also must follow the guidelines as set forth in other referenced standards. These include the following:

- Prefabricated wood I-joists—ASTM D 5055
- Structural glued-laminated timber—AITC A 190.1 and ASTM D 3737
- Wood structural panels—DOC PS 1 or PS 2
- Hardboard—AHA A 135.6
- Pre-finished hardboard—AHA A 135.5
- Particleboard—ANSI A208.1
- Preservative-treated wood—AWPA U1 and M4.

CODE UPDATE

Conventional light-frame construction is a type of construction where the primary structural elements are formed by a system of repetitive wood-framing members.

Did You Know?

When applied to the outside of foundation walls, fiberboard wall insulation must be protected below the ground level with a bituminous coating.

Fiberboard

Fiberboard and its various uses must conform to ASTM C 208, as well as fiberboard sheathing. Edges must be tight-fitting with square or U-shaped joints. When fiberboard is used as insulation, regardless of construction, you must ensure its protection with an approved roof covering.

Fiberboards are allowed as wall insulation in all types of construction as long as they are installed and fireblocked. If used in fire walls and barriers, but not treated to comply with Section 803.1 for Class A materials, you must be sure to cement the boards directly to the concrete or other noncombustible base. They must be protected with an approved noncombustible veneer anchored to the base without intervening airspaces.

Trusses

Wood used as trusses must be preservative treated and marked with a quality mark of an inspection agency that upholds continuing supervision, testing, and inspection over the quality of the preservative-treated wood. All inspection agencies for such wood must be listed by the accreditation body that complies with the requirements of the American Lumber Standards Treated Wood Program. The label that is affixed on the wood must contain the following:

- Identification of the manufacturer that treated the wood
- Type of preservative used
- Minimum preservative retention
- End use for which the product is treated
- AWPA standard to which the product was treated
- Identification of the inspection agency.

Did You Know?

Preservative-treated wood that is used in enclosed locations without a reasonable means of drying must have a moisture content of 19 percent or less before it can be covered with materials such as insulation, interior wall finish, or other materials.

Fire-Retardant-Treated Lumber

All fire-retardant treated lumber and wood structural panels also must be labeled with the following:

- An I.D. mark from an approved agency
- Identification of the treating manufacturer
- Name of the treatment
- Species of wood
- Flame spread and smoke index
- Method of drying that was used
- Conformance with appropriate standards
- For wood exposed to weather, include these words: "No increase in the listed classification when subjected to the Standard Rain Test" (ASTM D 2898).

Wherever fire-retardant treated wood is exposed to damp or wet locations, it must be identified as exterior. This indicates that there is no increase in the listed flame spread index. For interior uses, fire-retardant treated wood cannot have a moisture content over 28 percent when tested in accordance with ASTM D 3201. This type of wood has to be dried so that the moisture content is 19 percent or lower for lumber and 15 percent or lower for wood structural panels.

GENERAL CONSTRUCTION REQUIREMENTS

When working with general construction requirements, you will be working with lay-up patterns. The five standard patterns that can be used are as follows:

- Simple span pattern: All pieces are supported by two supports
- Two-span continuous pattern: All pieces are supported by three supports with all end joints occurring in line on every other support.
- Combination simple and two-span continuous pattern: End spans alternate with simple and two-span continuously. Stagger end joints in adjacent courses only over supports.
- Cantilevered pieces intermixed pattern: For this pattern, the decking covers a minimum of three spans. Pieces in the starter course and every third course must be simple span. Cantilever the pieces in the other course over the supports with end joints alternating.
- Controlled random pattern: This decking covers a minimum of three spans with end joints within 6 inches of being in line in either direction. There are no end joints in this pattern.

Be aware of changes for mechanically laminated decking if that is what you choose. This decking consists of square-edged dimension lumber laminations that are set on edge and nailed to the adjacent piece and to the support. Also, you cannot use nails that are less than 2.5 times the net thickness of each lamination either.

When using sheathing nails, make sure that the head or crown is just flush with the surface of the sheathing and not into the sheathing. The code does allow use of clips, glues, or any other methods as long as they are approved. When using fasteners in preservative-treated wood, you must use hot dipped zinc-coated galvanized steel. You can also use stainless steel or copper, but the coating weights must be in accordance with ASTM A 153.

CODE UPDATE

The following types of wood are considered termite resistant: redwood, Alaska yellow-cedar, eastern red cedar, and both heartwood and all sapwood of western red cedar.

The exception to this allows fasteners other than nails, timber rivets, or wood screws which are mechanically deposited zinc coated steel, but these must have a coating weight in accordance with ASTM B 695, Class 55 minimum.

As a reminder, fastenings for wood foundations must meet the requirements of AF&PA Technical Report No. 7.

Decay and Termites

Termite resistant wood must be used for wood floor framing, especially in areas heavily populated with termites. Water-borne preservatives, in accordance with AWPA U1, are used to preserve such wood for use above the ground. There are many woods used above the ground, such as wood joists, sleepers and sills, wood siding, and girder ends, which are connected in one way or another to the foundation.

GENERAL DESIGN REQUIREMENTS

All buildings and structures need to have lateral-force-resisting systems to resist seismic or wind loads. Then again, compliance with the AF&PA is allowed subject to the limitations of this code. To calculate shear resistance of diaphragms and shear walls, you can use principles of mechanics by using the values of fastener strength and sheathing shear resistance. You must have positive connections and anchorages between the shear panel and the attached components that are capable of resisting the design forces. The capacity of toenail connections cannot be used in Seismic Design Category D, E, or F when calculating lateral load resistance to transfer lateral earthquake forces in excess of 150 pounds per feet.

For structures of Seismic Design Category F, you must follow certain requirements. Wood structural panel sheathing that is used for diaphragms and shear walls that are part of the seismic-force-resisting systems must be

> **CODE UPDATE**
>
> Wood and wood-based materials that use vacuum-pressure impregnation processes to enhance fire retardant or preservative properties are known as treated wood.

CODE UPDATE

When hardboard siding is used structurally, it must be identified as a conforming material by an approved agency.

applied directly to the framing members, except for wood structural panel sheathing in a diaphragm which is allowed to be fastened over solid lumber planking only if the panel joints and lumber planking joints do not coincide.

When designing structures with rigid diaphragms, refer to the requirements of Section 12.3.2 of ASCE 7 and the horizontal shear distribution requirements of Section 12.8.4 of ASCE 7, as rigid diaphragms must comply with these sections. Refer to your local code book to find the diaphragm length and width for plan view of open-front buildings.

ALLOWABLE STRESS DESIGN

The following list of applicable standards contains the provisions in which the allowable stress design of wood elements must comply.

- American Forest & Paper Association
- American Institute of Timber Construction
- American Society of Agricultural Engineers
- APA—The Engineered Wood Association
- Truss Plate Institute, Inc.

There has been a change in the code for some parts of this section, such as wood diaphragms. Wood structural panel diaphragms are allowed to resist horizontal forces by using the allowable shear capacities. The allowable capacities that are found in the code must be increased by 40 percent for wind design, and sheathed lumber (diagonally) diaphragms must be nailed in accordance with nailing requirements.

This brings us to the application of gypsum board, or lath, and plaster to wood framing. There are several steps to take to ensure proper adherence. When joint staggering, make sure the end joints of adjacent courses of gypsum board do not occur over the same studs. You must be familiar with the maximum allowable shear values for blocking, fasteners, gypsum lath, and sheathing for gypsum boards.

CONVENTIONAL LIGHT-FRAME CONSTRUCTION

As the title suggests, this section is intended for conventional light-frame construction. Other methods are allowed, but only if you submit a satisfactory design that is in compliance in every way necessary with this code. One- and two-family homes and multiple single-family buildings, such as townhomes, no more than three stories above grade plane with separate means of exit, must comply with the International Residential Code.

Floor Joists

The spans for floor joists must comply with tables found in your local code book.

For other spans and species that you cannot locate in the building code, please refer to AF&PA Span Tables for Joists and Rafters. The ends of each joist, other than where supported on a 1 inch by 4 inch ribbon strip and nailed, cannot have less than 1½ inches of bearing on wood or metal or less than 3 inches. Make sure that all joists are supported laterally at the ends and at each support by solid blocking, except where the ends of the joists are nailed to the header.

The ends of header joints that are longer than 6 feet must be supported by framing anchors or joist hangers unless the header joints are bearing on a beam or wall.

A braced wall line is a series of braced wall panels in a single story. All braced wall lines must meet all the requirements of this code. Such requirements include location, type and amount of bracing.

!Definitionalert

Floor joists: The main sub-floor framing members that support the floor span.

Did You Know?

When joist framing is on opposite sides of a beam or girder, make sure it is lapped at least 3 inches. If not lapped, then tie the opposite ends together in the manner which has been approved. A joist must be cut away to give way for floor openings. Specifications usually require that headers be double and framed between the full-length joists, also known as trimmers, on either side of the floor opening.

Braced Wall Panels

There are several methods to construct braced wall panels. Use one of the following:

- Wood boards of ⅝-inch minimum thickness applied diagonally on studs with spacing of no more than 24 inches
- Fiberboard sheathing panels of no less than ½-inch thickness applied vertically or horizontally on studs with spacing of no more than 16 inches
- Portland cement plaster on studs spaced 16 inches

Remember that ceiling joists must have a minimum allowable span.

Purlins

Purlins that are installed for roof load support can be used to reduce the span of rafters within the allowable limits. These must be supported by struts to bearing walls and under no circumstance can a purlin be smaller than the rafter that it is supporting. This horizontal structural member must adhere to the allowable limits. For example, a purlin that is 2 inches by 4 inches, has a maximum length of 4 feet, and if the purlin is 2 inches by 6 inches, the maximum length is 6 feet.

Engineered Wood Products

As mentioned several times previously, there are many changes to the International Building Code. One of these changes states that engineered wood products, which can be composite lumber, prefabricated I-Joists, and struc-

!Definitionalert

Purlins: Framing members that support a roof panel assembly.

tural glue-laminated timber, cannot be notched or drilled unless the manufacturer has given permission to do so or unless the effects of doing so are specifically stated in the design by a registered design professional.

I make it a point to mention that there are additional requirements for conventional construction in Seismic Design Category B or C. In Seismic Design Category C, structures of conventional light-frame construction cannot be more than two stories in height, and concrete or masonry walls cannot extend above the basement. Be aware of the following exceptions:

- You may use masonry veneer in the first two stories above grade plane or in the first three stories where the lowest story has concrete or masonry for Category B, only if panel wall bracing is used and is 1½ times the required length.
- You may use masonry veneer in the first story above grade plane or the first two stories above grade plane where the lowest story has concrete or masonry walls in Category B or C.
- You may use masonry veneer in the first two stories above grade plane in Category B or C if the type of brace used in Section 2308.9.3 is Method 3 and the bracing of the top story is located at each end and at least every 25 feet. You must also provide hold-down connectors at the ends of braced walls for the second floor to the first floor assembly with a design of 2000 pounds.

CODE UPDATE

The written, graphic, and pictorial depiction of each individual truss shall be provided to the building official for approval prior to installation.

- You may not use cripple walls.

In the remainder of this chapter I will cover irregular structures. You may not use conventional light-frame construction in irregular parts of structures in Seismic Design Category D or E.

There are six conditions in which a portion of a building or structure will be deemed irregular. If your building or structure contains one or more of these conditions you have an irregular structure. Such conditions are as follows:

- Exterior braced wall panels are required to be in one plane vertically from the foundation to the uppermost story. If this is not true of your structure, it will be considered to be irregular. As with many sections of this book, this too has exceptions. Floors with cantilevers that are not more than four times the depth of the floor joists are permitted provided that the following is true:
  - Floor joists are 2 inches by 10 inches or larger.
  - The back span to the cantilever has a ratio of 2:1.
  - Floor joists at the ends of the braced walls are doubled.
  - A continuous rim joint is connected to the ends of cantilevered joists.
- The ends of cantilevered joists cannot carry gravity loads from more than a single story.
- Any section of a floor or roof that is not laterally supported by braced wall lines on all edges will be considered irregular; however, portions of roofs or floors that do not support braced wall panels are allowed to extend up to 6 feet.

CODE UPDATE

Wood products that are impregnated with chemicals by a pressure process shall be processed in a closed vessel under pressure not less than 50 pounds per square inch gauge.

CODE UPDATE

The design for the transfer of loads and anchorage of each truss to the supporting structure is the responsibility of the registered design professional.

- If the end of a required braced wall panel is more than 1 foot over an opening in the wall below, your structure is irregular. Braced wall panels are allowed to extend more than 8 feet in width if the header is 4 inches by 12 inches or larger.
- If portions of a floor level are vertically offset so the framing members on either side cannot be lapped or tied, you will qualify for an irregular structure. Framing supported directly by foundations do not need to be lapped or tied.
- If braced wall lines are not perpendicular to each other, you have an irregular structure.
- Openings in floor and roof diaphragms that have a maximum dimension more than 50 percent of the distance between lines of bracing will put your structure in the irregular structure category.

CODE UPDATE

Where wood joists or the bottom of a wood structural floor without joist are closer than 18 inches, or wood girder are closer than 12 inches to the exposed ground in crawl spaces or unexcavated areas located within the perimeter of the building foundation, the floor construction shall be of naturally durable or preservative-treated wood.

CHAPTER

13

Plastic

It seems that we use more and more plastic components as time moves forward. They are a part of construction that is not likely to go away anytime soon.

The types of plastics included in this section are as follows:

- Foam plastic
- Foam plastic insulation
- Plastic veneer
- Interior plastic finish and trim
- Light-transmitting plastics.

FOAM PLASTIC INSULATION

When your package of foam plastic insulation arrives at the job site, see that the label of an approved agency is attached to the package or the container and shows the manufacturer's name, the product listing, the product identification and any information that will be used to identify the end

CODE UPDATE

A polymeric composite material consisting of reinforcement fibers, impregnated with a fiber-binding polymer which is then molded and hardened is a fiber reinforced polymer.

use with the code requirements. Both foam plastic insulation and foam plastic cores of manufactured assemblies will have a flame spread index of not more than 75. A smoke-developed index of not more than 450 must also be a characteristic of the foam insulation, and both must be in accordance with ASTM E 84. All loose fill-type foam plastic must be tested as board stock for the flame spread index and smoke-developed index. A list of exceptions is as follows:

- Section 2604.2 contains the smoke-developed index for interior trim.
- In buildings that are used as cold storage, like ice plants, food plants, or food processing rooms, where insulation is tested in a thickness of 4 inches, insulation will be allowed in a thickness up to 10 inches if the building is equipped throughout with an automatic fire sprinkler.
- Roof-covering assemblies such as Class A, B, or C that contain foam plastic insulation provided that the insulation passes FM 4450 or UL 1256
- Foam plastic insulation that is greater than 4 inches in thickness must have a maximum flame index of 450 when tested at a minimum thickness of 4 inches.
- Flame spread and smoke-developed indexes for foam-developed plastic interior signs in covered mall buildings provided the signs comply with Section 402.15.

With the exceptions of masonry and concrete construction, and cooler and freezer walls, foam plastic must be separated from the inside of the building by a thermal barrier of ½-inch gypsum wallboard. This is to limit the average temperature rise of the unexposed surface to no more than 250°F after 15 minutes of fire exposure, complying with the standard time-temperature curve of ASTM E 119. As stated previously, no thermal barrier is needed in masonry or concrete construction. This is true if the concrete wall, floor, or roof system is covered on each face by a minimum of 1-inch thickness of masonry or concrete. Nor is a thermal barrier required for cooler and freezer walls that are installed with foam plastic with a thickness of 10 inches. Cooler and freezer walls must:

- Have a flame spread index of 25 or less with a smoke-developed index of no more than 450
- Have flash ignition and self-ignition temperatures of no less than 600°F to 800°F
- Have a covering of no less than 0.032-inch corrosion-resistant steel with a base metal thickness no less than 0.0160 inch at any point
- Be protected by an automatic sprinkler system
- A cooler or freezer within a building must also be part of the building protected by sprinklers.

In one-story buildings that have a foam plastic with a flame spread index of less than 25, cooler and freezer walls are not required to have a thermal barrier in or on outside walls in a thickness no more than 4 inches thick. This is only true where the foam is covered by a no less than 0.032-inch-thick aluminum steel with a base thickness of 0.0160 inch. The building must be fully equipped throughout with an automatic sprinkler system. Foam plastic insulation mounted in accordance with both the code and the manufacturer's instructions must be separated from the inside of the building with a wood panel sheathing no less than 0.47 inch thick and bonded with exterior glue.

The following list contains the provisions for doors that are not required to have a fire protection rating:

- Pivoted or side-hinged doors having a flame spread index of 75 or less are allowed as a core material.
- For exterior doors in buildings of Group R-2 or R-3, foam-filled entrance doors must be faced with wood or another approved material.

!Definitionalert

Foam plastic insulation: A plastic that expands with the use of a foaming agent producing a reduced-density plastic with voids of open or closed cells distributed throughout the plastic for thermal insulating purposes. Foam plastic insulation has a density less than 20 pounds pcf.

- Garage doors with foam plastic used as core materials must have a door facing of metal with a thickness of 0.032 inch.

There is an exception for garage doors using foam plastic insulation and the compliance with Section 2603.3 in regards to detached and attached garages of one- and two-family dwellings that exempt them from the thermal barrier requirement. Anytime that foam plastic is used as interior trim, and compliance with Section 2604 is met, a thermal barrier is not needed. This also is true for interior signs that comply with Section 402.15. However, be aware that foam plastic signs that are not attached to the interior of a building must comply with the International Fire Code.

Foam plastic insulation, exterior coatings and facings all must be tested separately for the thickness intended for use, flame index spreads, and smoke-developed index. ASTM E 84 has determined that the flame index spread for all three cannot be more than 25 with a smoke-developed index of less than 450. However, prefabricated or factory-manufactured panels

Did You Know?

When you apply foam plastic spray to a sill plate and header of Type V construction there are three requirements that must be met. Plastic foam spray must have a maximum thickness of 3¼ inches, foam plastic must have a density range of 1.5 to 2.0 pcf, a flame spread index of 25 or less, and a smoke-developed index of 450 or less when tested with ASTM E 84.

can be tested as an assembly as long as the foam plastic core is not exposed during construction and only if panels have a minimum thickness of 0.020-inch aluminum facing and a total thickness of 0.25 inch or less.

There has been a change in this code regarding the use of foam plastics and the protection against termites.

It is now known that termites have a huge tendency to burrow through expanded polystyrene and other foam plastics to get to wood. Therefore, you cannot install exterior foam plastics below grade. There are a few exceptions as follows:

- In buildings where walls, floors, and ceilings are made of wood treated with a preservative or noncombustible materials
- When there has been an approved method used to protect foam plastic and structures from termite damage
- On the inside of basement walls.

Foam plastics are not required to comply with Sections 2603.4 through 2603.7 if approval has been met through testing. This testing is not limited to NFPA 286, FM 4880, UL 140, or 1715, and any testing must be in relation to the use of the foam plastic. When allowed for use as an interior finish—and only on the basis of special testing will this be allowed. Testing will include seams, joints, and other details typically used for this type of installation. Testing must also include the manner intended for use.

INTERIOR FINISH AND TRIM

Plastic materials must comply with this section when installed as interior finish and trim. Also refer to the last sentence of the above paragraph as a reminder of special approval. When and if you receive special approval to use plastic materials for trim you must also comply with the following:

- Interior trim must have a minimum density of 20 pcf.
- Interior trim must have a maximum thickness of ½ inch with a maximum width of 8 inches.
- There will be an area limitation of no more than 10 percent of the aggregate wall and ceiling area of any room or space.

- When tested in accordance with ASTM E 84, the flame spread index will not be more than 75.
- No limitation is made on the smoke-developed index.

PLASTIC VENEER

Plastic veneer used inside of a building must be in compliance with interior finish and trim. You are allowed to install exterior plastic veneer on the exterior walls of buildings and are not restricted to the type of construction; however, there are guidelines that must be met. See the following list for these guidelines:

- Plastic veneer must comply with Section 2606.4.
- Do not attach plastic veneer to any outside wall with a height of more than 50 feet above grade.
- There is a 300 square foot minimum for sections of plastic veneer with a separation minimum of 4 feet.
- Note that the area and separation requirements and the smoke-density limitation do not apply to Type VB construction, provided that there is not a fire-resistance rating requirement for walls.

LIGHT-TRANSMITTING PLASTICS

Before any light-transmitting materials are used, you must first submit, for approval, sufficient data to demonstrate the use of such materials. It is up to the building official to allow the use of light-transmitting materials in your building or structure. Light-transmitting plastics include thermoplastic, thermosetting, or reinforced thermosetting plastic materials, and all of these must have a self-ignition temperature of 650°F or higher when tested in accordance with ASTM D 1929.

CODE UPDATE

Foam plastics installed as trim must not constitute more than 10 percent of the specific wall or ceiling areas to which it is attached.

The smoke-developed index cannot be greater than 450 when tested in accordance with ASTM E 84, or greater than 75 when tested in accordance with ASTM D 2843. All light-transmitting materials must conform to either Class CC1, in which plastic materials have a 1 inch or less burning extent when tested at a thickness of 0.060 inch, or Class CC2, in which plastic materials have a burning rate of 2½ inches per minute when tested at a thickness of 0.060 inch. Both classes must be tested in accordance with ASTM D 635.

Light-transmitting plastic materials must comply with provisions regarding the durability of strength and loads. The building official will use any data that establishes stresses, maximum weight spans, and other important information in making decisions regarding the use of light-transmitting plastics. Submission of this information and data is mandatory on your part; without doing so, you will not have approval from the building official.

There are certain occupancies and locations in which a light-diffusing system cannot be installed without an automatic sprinkler system. These occupancies and locations are as follows:

- Group A occupancies with an occupant load of 1000 or more
- Theaters with a stage and proscenium opening and an occupant load of more than 700
- Group I-2 and I-3 occupancies
- Stairways and passageways intended for exits.

Hangers of at least No. 12 steel must be used as support for light-transmitting diffusers. These must be hung directly or indirectly from ceilings or roof construction and the hangers must be made of a noncombustible material. Be sure that you are in compliance with the code, unless the plastic diffusers have the ability to fall from their mountings before ignition takes place at a temperature of 200°F; this temperature is below the ignition temperature of panels. Please note that panels cannot be more than 10 feet in length or 30 square feet in length.

You are allowed, by this code, to install light-transmitting plastics as glazing in shower stalls, doors, and bathtub enclosures as long as you are in compliance with Glass and Glazing.

LIGHT-TRANSMITTING PLASTIC WALL PANELS

You cannot use light-transmitting plastic as wall panels in exterior walls in occupancies of the following groups.

- A-1
- A-2
- H
- I-2
- I-3.

This is allowed in other groups, but only if the walls are not required to have a fire-resistance rating, and installation conforms to this section. For instance, exterior wall panels cannot alter the type of construction.

Lastly, the combining of light-transmitting plastic glazing and light-transmitting plastic wall panels is subject to height, area, and percentage limitations along with any separation requirements.

LIGHT-TRANSMITTING PLASTIC GLAZING

Be reminded that there are provisions regarding the protection for openings in exterior walls of Type VB construction. However, areas that are not required to comply are permitted to have light-transmitting plastic as long as it in compliance with its proper section of this chapter and also in accordance with the following.

- The total area of the plastic glazing cannot be more than 25 percent of any area of any wall face of the story in which it is installed, except for those areas that have an automatic sprinkler system; only then can the total area be increased to a maximum of 50 percent.
- Installation of an approved flame barrier between glazed units in adjacent stories is allowed and it must extend 30 inches beyond the outside wall or a vertical panel no less than 4 feet. An exception to this is a building equipped throughout with an automatic sprinkler system.

- Do not install light-transmitting plastics more than 75 feet above grade plane. Again, an exception to this is a building equipped throughout with an automatic sprinkler system.

LIGHT-TRANSMITTING PLASTIC ROOF PANELS

Such panels cannot be installed in Groups H, I-2, and I-3. You may install them in other groups as long as they comply with any one of the following conditions.

- The building has an automatic sprinkler system throughout.
- The roof is not required to comply with the fire-resistance rating
- Roof panels meet the requirements for roof coverings.

Roof panels are required to be limited in area, and the total area of panels is limited by a percentage of the floor area. The following list of exceptions address this provision.

- Area limitations are increased by 100 percent if the building has an automatic sprinkler system.
- Low-hazard occupancy buildings are exempt from limitations provided that the buildings are no larger than 5000 square feet with a minimum fire separation distance of 10 feet.
- Greenhouses without public access are exempt from area limitations, but only if they have a minimum fire separation distance of 4 feet.

CODE UPDATE

Light-transmitting plastic covers on solar collectors having noncombustible sides and bottoms shall be permitted on buildings not over three stories above grade plane of 9000 square feet in total floor area, provided the light-transmitting plastic cover does not exceed one third of the roof area for CC1 materials or 25 percent of the roof area for CC2 materials.

- Roof coverings of terraces and patios in Group R-3 occupancies are permitted to have light-transmitting plastics.

LIGHT-TRANSMITTING PLASTIC SKYLIGHT GLAZING

This section provides the provisions for skylight assemblies with light-transmitting plastics. If installing unit skylights glazed with light-transmitting plastic, refer to the code book for compliance requirements. This provision does not include skylights that contain light-transmitting plastics that already conform to code. It is ruled that light-transmitting plastics be set upon curbs that have been constructed above the plane of the roof by at least 4 inches. Make sure that all edges are protected by metal or another material that has been approved for protective use.

Any material you use must be ignition resistant to exposure of flames. Curbs are not required for skylights that are used on roofs with a minimum slope of three units vertical in 12 units horizontal. This pertains to occupancies in Group R-3 and for buildings that have a noncombustible roof covering. Also, metal or noncombustible edge materials are not required where non-classified roof coverings are allowed.

There is a maximum of 100 square feet within the curb area that skylights must have; however, this area limitation does not apply to buildings equipped with an automatic sprinkler system. Make sure that you have placed all your skylights with a distance of no less than 4 feet measured in a horizontal plane.

Buildings with automatic sprinkler systems and Group R-3 occupancies with more than one skylight that do not exceed the area limitations in this section are not included in the separation provisions.

LIGHT-TRANSMITTING PLASTIC INTERIOR SIGNS

Light-transmitting plastic interior signs must follow the provisions of the code and cannot have an area more than 20 percent of the wall area with a maximum area of 24 square feet. The edges and backs of all plastic interior signs must be fully encased in metal.

CHAPTER

14

Encroachments Into the Public Right-of-Way

Due to the lack of pertinent information, this book does not include commentary regarding encroachments into the public right-of-way. However, this is an important area of reference in the 2009 International Building Code.

APPENDIX

Reference Material

TYPE OF MATERIAL	THICKNESS OF MATERIAL (Inches)											
	1 1/2	2	2 1/2	3	3 1/2	4	4 1/2	5	5 1/2	6	6 1/2	7
Siliceous aggregate concrete	5.3	6.5	8.1	9.5	11.3	13.0	14.9	16.9	18.8	20.7	22.8	25.1
Carbonate aggregate concrete	5.5	7.1	8.9	10.4	12.0	14.0	16.2	18.1	20.3	21.9	24.7	27.2[c]
Sand-lightweight concrete	6.5	8.2	10.5	12.8	15.5	18.1	20.7	23.3	26.0[c]	Note c	Note c	Note c
Lightweight concrete	6.6	8.8	11.2	13.7	16.5	19.1	21.9	24.7	27.8[c]	Note c	Note c	Note c
Insulating concrete[a]	9.3	13.3	16.6	18.3	23.1	26.5[c]	Note c	Note c	Note c	Note c	Note c	Note c
Airspace[b]	—	—	—	—	—	—	—	—	—	—	—	—

For SI: 1 inch = 25.4 mm, 1 pound per cubic foot = 16.02 kg/m³.

a. Dry unit weight of 35 pcf or less and consisting of cellular, perlite or vermiculite concrete.

b. The $R_n^{0.59}$ value for one 1/2" to 3 1/2" airspace is 3.3. The $R_n^{0.59}$ value for two 1/2" to 3 1/2" airspaces is 6.7.

c. The fire-resistance rating for this thickness exceeds 4 hours.

APPENDIX FIGURE 1

Values of $R_n^{0.59}$.

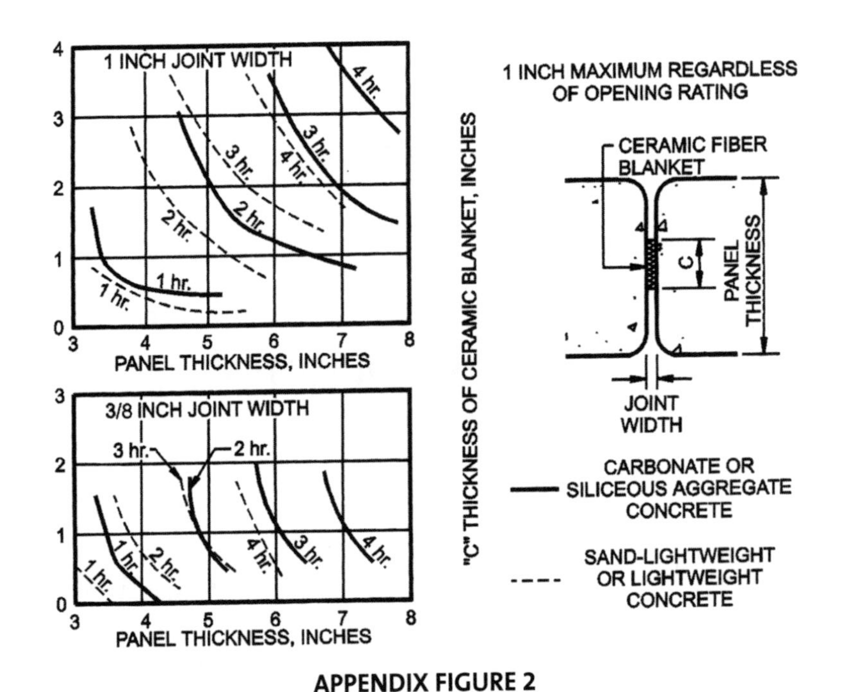

APPENDIX FIGURE 2

Ceramic fiber joint protection.

FINISH DESCRIPTION	TIME (minute)
Gypsum wallboard	
$^3/_8$ inch	10
$^1/_2$ inch	15
$^5/_8$ inch	20
2 layers of $^3/_8$ inch	25
1 layer $^3/_8$ inch, 1 layer $^1/_2$ inch	35
2 layers $^1/_2$ inch	40
Type X gypsum wallboard	
$^1/_2$ inch	25
$^5/_8$ inch	40
Portland cement-sand plaster applied directly to concrete masonry	See Note a
Portland cement-sand plaster on metal lath	
$^3/_4$ inch	20
$^7/_8$ inch	25
1 inch	30
Gypsum sand plaster on $^3/_8$-inch gypsum lath	
$^1/_2$ inch	35
$^5/_8$ inch	40
$^3/_4$ inch	50
Gypsum sand plaster on metal lath	
$^3/_4$ inch	50
$^7/_8$ inch	60
1 inch	80

For SI: 1 inch = 25.4 mm.

a. The actual thickness of portland cement-sand plaster, provided it is $^5/_8$ inch or less in thickness, shall be permitted to be included in determining the equivalent thickness of the masonry for use in Table 721.3.2.

APPENDIX FIGURE 3

Time assigned to finish materials on fire-exposed side of wall.

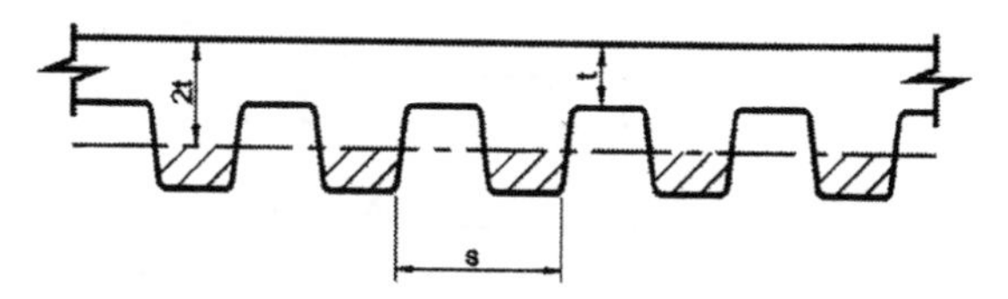

NEGLECT SHADED AREA IN CALCULATION OF EQUIVALENT THICKNESS

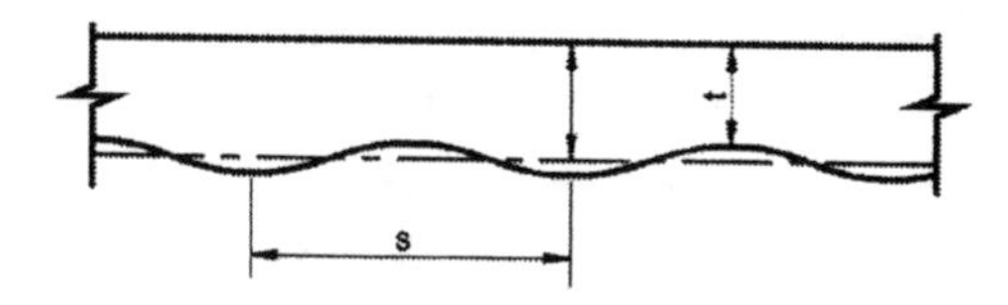

For SI: 1 inch = 25.4 mm.

APPENDIX FIGURE 4

Slabs with ribbed or undulating soffits.

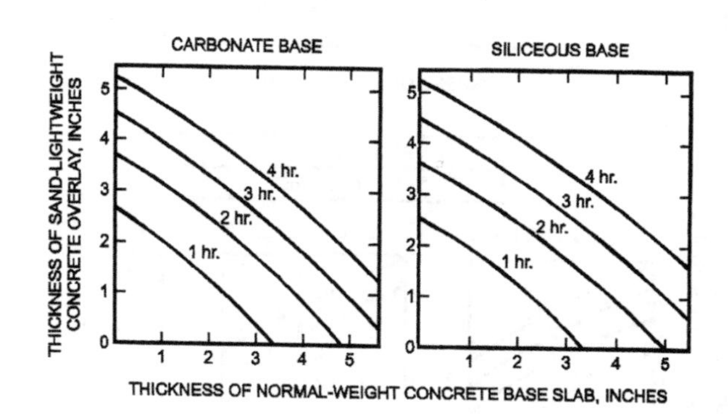

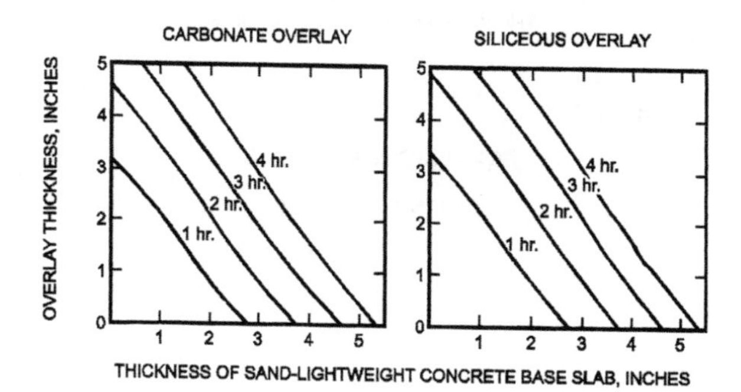

For SI: 1 inch = 25.4 mm.

APPENDIX FIGURE 5

Fire-resistance ratings for two-course concrete floors.

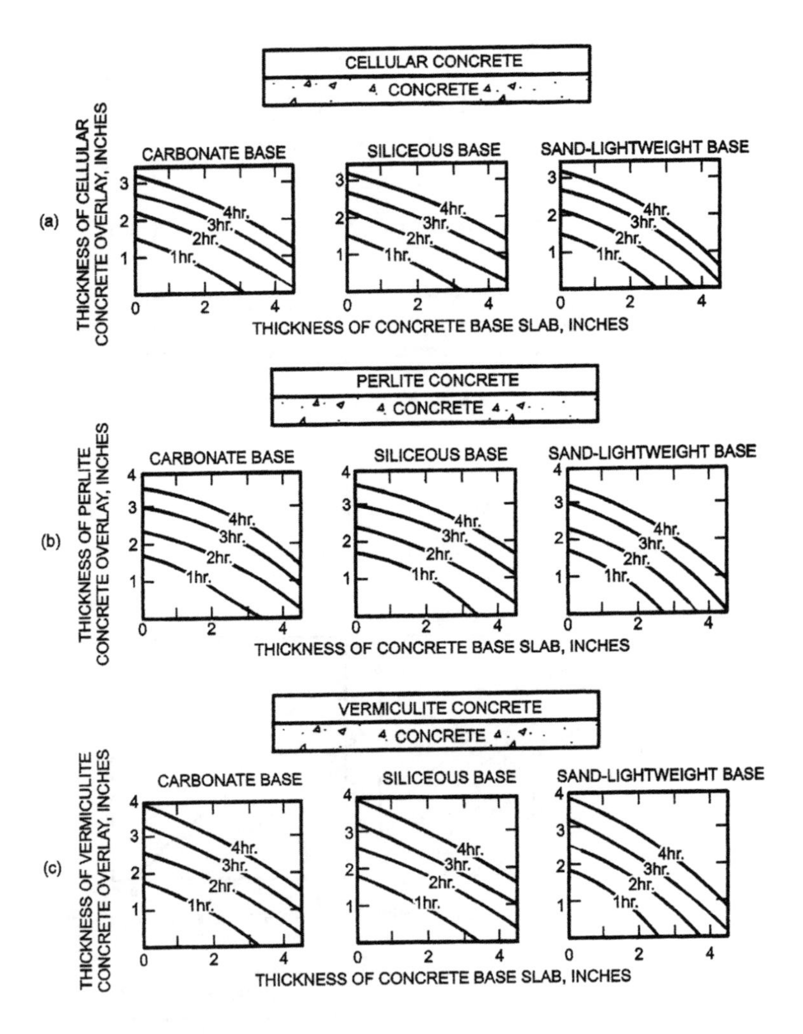

For SI: 1 inch = 25.4 mm.

APPENDIX FIGURE 6

Fire-resistance ratings for concrete roof assemblies.

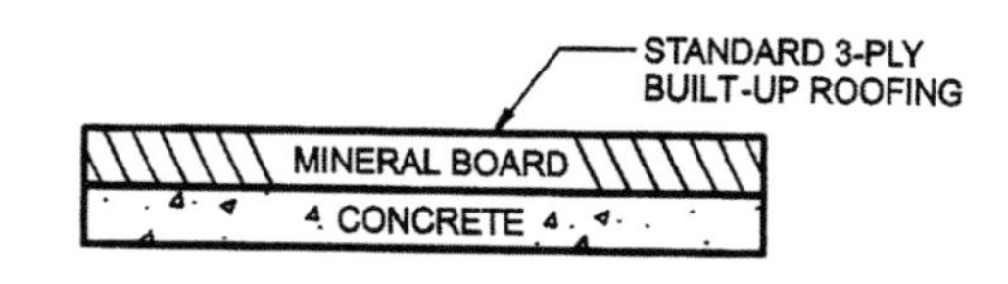

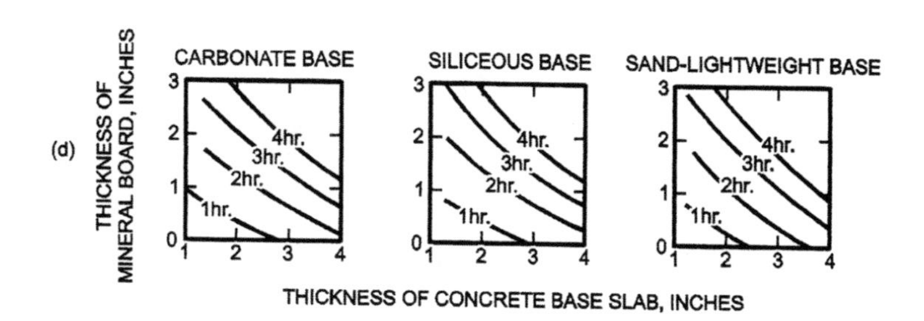

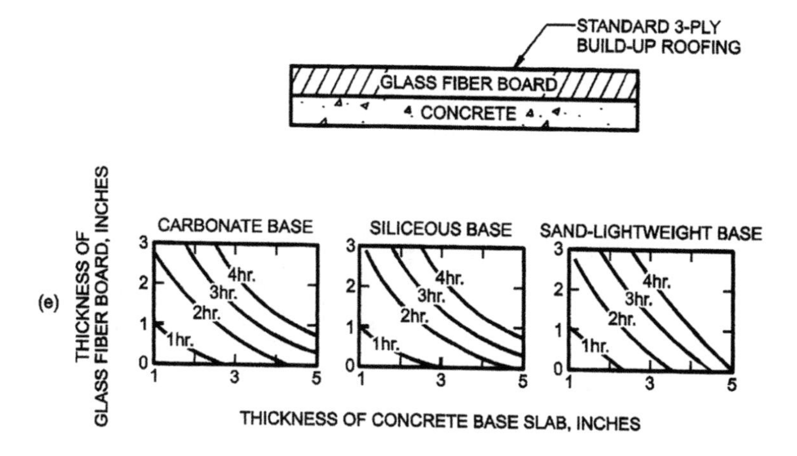

For SI: 1 inch = 25.4 mm.

APPENDIX FIGURE 7

Fire-resistance ratings for concrete roof assemblies.

CONCRETE AGGREGATE TYPE	FIRE-RESISTANCE RATING (hours)									
	Restrained					Unrestrained				
	1	$1^1/_2$	2	3	4	1	$1^1/_2$	2	3	4
Siliceous	$^3/_4$	$^3/_4$	$^3/_4$	$^3/_4$	$^3/_4$	$^3/_4$	$^3/_4$	1	$1^1/_4$	$1^5/_8$
Carbonate	$^3/_4$	$^3/_4$	$^3/_4$	$^3/_4$	$^3/_4$	$^3/_4$	$^3/_4$	$^3/_4$	$1^1/_4$	$1^1/_4$
Sand-lightweight or lightweight	$^3/_4$	$^3/_4$	$^3/_4$	$^3/_4$	$^3/_4$	$^3/_4$	$^3/_4$	$^3/_4$	$1^1/_4$	$1^1/_4$

For SI: 1 inch = 25.4 mm.

APPENDIX FIGURE 8

Cover thickness for reinforced concrete floor or roof slabs (inches).

CONCRETE AGGREGATE TYPE	FIRE-RESISTANCE RATING (hours)									
	Restrained					Unrestrained				
	1	$1^1/_2$	2	3	4	1	$1^1/_2$	2	3	4
Siliceous	$^3/_4$	$^3/_4$	$^3/_4$	$^3/_4$	$^3/_4$	$1^1/_8$	$1^1/_2$	$1^3/_4$	$2^3/_8$	$2^3/_4$
Carbonate	$^3/_4$	$^3/_4$	$^3/_4$	$^3/_4$	$^3/_4$	1	$1^3/_8$	$1^5/_8$	$2^1/_8$	$2^1/_4$
Sand-lightweight or lightweight	$^3/_4$	$^3/_4$	$^3/_4$	$^3/_4$	$^3/_4$	1	$1^3/_8$	$1^1/_2$	2	$2^1/_4$

For SI: 1 inch = 25.4 mm.

APPENDIX FIGURE 9

Cover thickness for prestressed concrete floor or roof slabs (inches).

RESTRAINED OR UNRESTRAINED[a]	BEAM WIDTH[b] (inches)	FIRE-RESISTANCE RATING (hours)				
		1	$1^1/_2$	2	3	4
Restrained	5	$^3/_4$	$^3/_4$	$^3/_4$	1[a]	$1^1/_4$[a]
	7	$^3/_4$	$^3/_4$	$^3/_4$	$^3/_4$	$^3/_4$
	≥ 10	$^3/_4$	$^3/_4$	$^3/_4$	$^3/_4$	$^3/_4$
Unrestrained	5	$^3/_4$	1	$1^1/_4$	—	—
	7	$^3/_4$	$^3/_4$	$^3/_4$	$1^3/_4$	3
	≥ 10	$^3/_4$	$^3/_4$	$^3/_4$	1	$1^3/_4$

For SI: 1 inch = 25.4 mm, 1 foot = 304.8 mm.

a. Tabulated values for restrained assemblies apply to beams spaced more than 4 feet on center. For restrained beams spaced 4 feet or less on center, minimum cover of $^3/_4$ inch is adequate for ratings of 4 hours or less.

b. For beam widths between the tabulated values, the minimum cover thickness can be determined by direct interpolation.

c. The cover for an individual reinforcing bar is the minimum thickness of concrete between the surface of the bar and the fire-exposed surface of the beam. For beams in which several bars are used, the cover for corner bars used in the calculation shall be reduced to one-half of the actual value. The cover for an individual bar must be not less than one-half of the value given in Table 721.2.3(3) nor less than $^3/_4$ inch.

APPENDIX FIGURE 10

Minimum cover for main reinforcing bars of reinforced concrete beams (applicable to all types of structural concrete).

RESTRAINED OR UNRESTRAINED[a]	CONCRETE AGGREGATE TYPE	BEAM WIDTH[b] (inches)	FIRE-RESISTANCE RATING (hours)				
			1	$1\frac{1}{2}$	2	3	4
Restrained	Carbonate or siliceous	8	$1\frac{1}{2}$	$1\frac{1}{2}$	$1\frac{1}{2}$	$1\frac{3}{4}$[a]	$2\frac{1}{2}$[a]
	Carbonate or siliceous	≥ 12	$1\frac{1}{2}$	$1\frac{1}{2}$	$1\frac{1}{2}$	$1\frac{1}{2}$	$1\frac{7}{8}$[a]
	Sand lightweight	8	$1\frac{1}{2}$	$1\frac{1}{2}$	$1\frac{1}{2}$	$1\frac{1}{2}$	2[a]
	Sand lightweight	≥ 12	$1\frac{1}{2}$	$1\frac{1}{2}$	$1\frac{1}{2}$	$1\frac{1}{2}$	$1\frac{5}{8}$[a]
Unrestrained	Carbonate or siliceous	8	$1\frac{1}{2}$	$1\frac{3}{4}$	$2\frac{1}{2}$	5[c]	—
	Carbonate or siliceous	≥ 12	$1\frac{1}{2}$	$1\frac{1}{2}$	$1\frac{7}{8}$[a]	$2\frac{1}{2}$	3
	Sand lightweight	8	$1\frac{1}{2}$	$1\frac{1}{2}$	2	$3\frac{1}{4}$	—
	Sand lightweight	≥ 12	$1\frac{1}{2}$	$1\frac{1}{2}$	$1\frac{5}{8}$	2	$2\frac{1}{2}$

For SI: 1 inch = 25.4 mm, 1 foot = 304.8 mm.

a. Tabulated values for restrained assemblies apply to beams spaced more than 4 feet on center. For restrained beams spaced 4 feet or less on center, minimum cover of $\frac{3}{4}$ inch is adequate for 4-hour ratings or less.

b. For beam widths between 8 inches and 12 inches, minimum cover thickness can be determined by direct interpolation.

c. Not practical for 8-inch-wide beam but shown for purposes of interpolation.

APPENDIX FIGURE 11

Minimum cover for prestressed concrete beams 8 inches or greater in width.

RESTRAINED OR UNRESTRAINED[a]	CONCRETE AGGREGATE TYPE	BEAM AREA[b] A (square inches)	FIRE-RESISTANCE RATING (hours)				
			1	$1^1/_2$	2	3	4
Restrained	All	$40 \le A \le 150$	$1^1/_2$	$1^1/_2$	2	$2^1/_2$	—
	Carbonate or	$150 < A \le 300$	$1^1/_2$	$1^1/_2$	$1^1/_2$	$1^3/_4$	$2^1/_2$
	siliceous	$300 < A$	$1^1/_2$	$1^1/_2$	$1^1/_2$	$1^1/_2$	2
	Sand lightweight	$150 < A$	$1^1/_2$	$1^1/_2$	$1^1/_2$	$1^1/_2$	2
Unrestrained	All	$40 \le A \le 150$	2	$2^1/_2$	—	—	—
	Carbonate or	$150 < A \le 300$	$1^1/_2$	$1^3/_4$	$2^1/_2$	—	—
	siliceous	$300 < A$	$1^1/_2$	$1^1/_2$	2	3[c]	4[c]
	Sand lightweight	$150 < A$	$1^1/_2$	$1^1/_2$	2	3[c]	4[c]

For SI: 1 inch = 25.4 mm, 1 foot = 304.8 mm.

a. Tabulated values for restrained assemblies apply to beams spaced more than 4 feet on center. For restrained beams spaced 4 feet or less on center, minimum cover of $^3/_4$ inch is adequate for 4-hour ratings or less.

b. The cross-sectional area of a stem is permitted to include a portion of the area in the flange, provided the width of the flange used in the calculation does not exceed three times the average width of the stem.

c. U-shaped or hooped stirrups spaced not to exceed the depth of the member and having a minimum cover of 1 inch shall be provided.

APPENDIX FIGURE 12

Minimum cover for prestressed concrete beams of all widths.

MATERIAL TYPE	MINIMUM REQUIRED EQUIVALENT THICKNESS FOR FIRE RESISTANCE[a, b, c] (inches)			
	1 hour	2 hour	3 hour	4 hour
Solid brick of clay or shale[d]	2.7	3.8	4.9	6.0
Hollow brick or tile of clay or shale, unfilled	2.3	3.4	4.3	5.0
Hollow brick or tile of clay or shale, grouted or filled with materials specified in Section 721.4.1.1.3	3.0	4.4	5.5	6.6

For SI: 1 inch = 25.4 mm.

a. Equivalent thickness as determined from Section 721.4.1.1.

b. Calculated fire resistance between the hourly increments listed shall be determined by linear interpolation.

c. Where combustible members are framed in the wall, the thickness of solid material between the end of each member and the opposite face of the wall, or between members set in from opposite sides, shall not be less than 93 percent of the thickness shown.

d. For units in which the net cross-sectional area of cored brick in any plane parallel to the surface containing the cores is at least 75 percent of the gross cross-sectional area measured in the same plane.

APPENDIX FIGURE 13

Fire-resistance periods of clay masonry walls.

WALL OR PARTITION ASSEMBLY	PLASTER SIDE EXPOSED (hours)	BRICK FACED SIDE EXPOSED (hours)
Outside facing of steel studs: $^1/_2$" wood fiberboard sheathing next to studs, $^3/_4$" airspace formed with $^3/_4$" × 1 $^5/_8$" wood strips placed over the fiberboard and secured to the studs; metal or wire lath nailed to such strips, 3$^3/_4$" brick veneer held in place by filling $^3/_4$" airspace between the brick and lath with mortar. Inside facing of studs: $^3/_4$" unsanded gypsum plaster on metal or wire lath attached to $^5/_{16}$" wood strips secured to edges of the studs.	1.5	4
Outside facing of steel studs: 1" insulation board sheathing attached to studs, 1" airspace, and 3$^3/_4$" brick veneer attached to steel frame with metal ties every 5th course. Inside facing of studs: $^7/_8$" sanded gypsum plaster (1:2 mix) applied on metal or wire lath attached directly to the studs.	1.5	4
Same as above except use $^7/_8$" vermiculite—gypsum plaster or 1" sanded gypsum plaster (1:2 mix) applied to metal or wire.	2	4
Outside facing of steel studs: $^1/_2$" gypsum sheathing board, attached to studs, and 3$^3/_4$" brick veneer attached to steel frame with metal ties every 5th course. Inside facing of studs: $^1/_2$" sanded gypsum plaster (1:2 mix) applied to $^1/_2$" perforated gypsum lath securely attached to studs and having strips of metal lath 3 inches wide applied to all horizontal joints of gypsum lath.	2	4

For SI: 1 inch = 25.4 mm.

APPENDIX FIGURE 14

Fire-resistance ratings for bearing steel frame brick veneer walls or partitions.

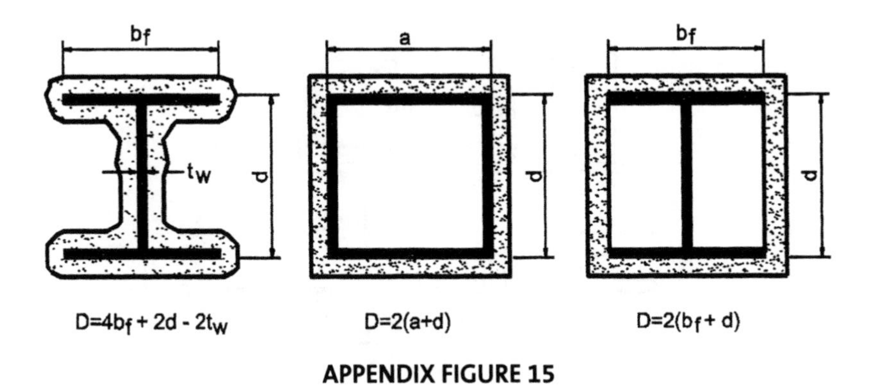

APPENDIX FIGURE 15

Determination of the heated perimeter of structural steel columns.

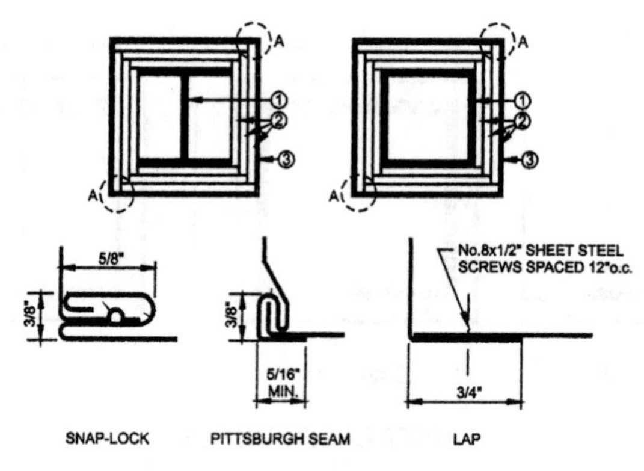

For SI: 1 inch = 25.4 mm, 1 foot = 305 mm.

1. Structural steel column, either wide flange or tubular shapes.
2. Type X gypsum wallboard in accordance with ASTM C 36. For single-layer applications, the wallboard shall be applied vertically with no horizontal joints. For multiple-layer applications, horizontal joints are permitted at a minimum spacing of 8 feet, provided that the joints in successive layers are staggered at least 12 inches. The total required thickness of wallboard shall be determined on the basis of the specified fire-resistance rating and the weight-to-heated-perimeter ratio (*W/D*) of the column. For fire-resistance ratings of 2 hours or less, one of the required layers of gypsum wallboard may be applied to the exterior of the sheet steel column covers with 1-inch-long Type S screws spaced 1 inch from the wallboard edge and 8 inches on center. For such installations, 0.0149-inch minimum thickness galvanized steel corner beads with $1^1/_2$-inch legs shall be attached to the wallboard with Type S screws spaced 12 inches on center.
3. For fire-resistance ratings of 3 hours or less, the column covers shall be fabricated from 0.0239-inch minimum thickness galvanized or stainless steel. For 4-hour fire-resistance ratings, the column covers shall be fabricated from 0.0239-inch minimum thickness stainless steel. The column covers shall be erected with the Snap Lock or Pittsburgh joint details.

   For fire-resistance ratings of 2 hours or less, column covers fabricated from 0.0269-inch minimum thickness galvanized or stainless steel shall be permitted to be erected with lap joints. The lap joints shall be permitted to be located anywhere around the perimeter of the column cover. The lap joints shall be secured with $^1/_2$-inch-long No. 8 sheet metal screws spaced 12 inches on center.

   The column covers shall be provided with a minimum expansion clearance of $^1/_8$ inch per linear foot between the ends of the cover and any restraining construction.

APPENDIX FIGURE 16

Gypsum wallboard protected structural steel columns with sheet steel column covers.

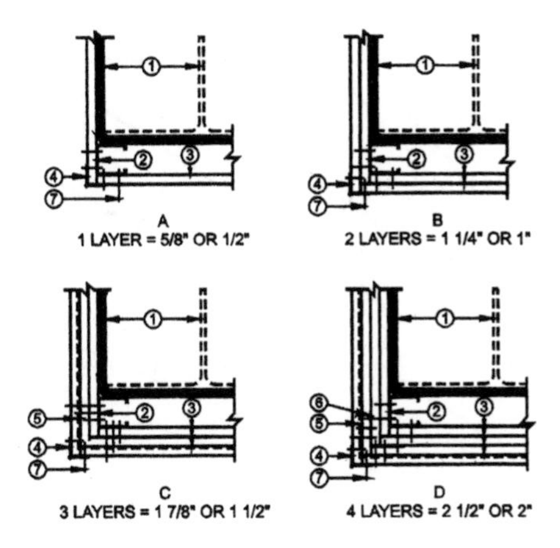

For SI: 1 inch = 25.4 mm, 1 foot = -305 mm.

1. Structural steel column, either wide flange or tubular shapes.
2. $1^5/_8$-inch deep studs fabricated from 0.0179-inch minimum thickness galvanized steel with $1^5/_{16}$ or $1^7/_{16}$-inch legs. The length of the steel studs shall be $^1/_2$ inch less than the height of the assembly.
3. Type X gypsum wallboard in accordance with ASTM C 36. For single-layer applications, the wallboard shall be applied vertically with no horizontal joints. For multiple-layer applications, horizontal joints are permitted at a minimum spacing of 8 feet, provided that the joints in successive layers are staggered at least 12 inches. The total required thickness of wallboard shall be determined on the basis of the specified fire-resistance rating and the weight-to-heated-perimeter ratio (*W/D*) of the column.
4. Galvanized 0.0149-inch minimum thickness steel corner beads with $1^1/_2$-inch legs attached to the wallboard with 1-inch-long Type S screws spaced 12 inches on center.
5. No. 18 SWG steel tie wires spaced 24 inches on center.
6. Sheet metal angles with 2-inch legs fabricated from 0.0221-inch minimum thickness galvanized steel.
7. Type S screws, 1 inch long, shall be used for attaching the first layer of wallboard to the steel studs and the third layer to the sheet metal angles at 24 inches on center. Type S screws $1^3/_4$-inch long shall be used for attaching the second layer of wallboard to the steel studs and the fourth layer to the sheet metal angles at 12 inches on center. Type S screws $2^1/_4$ inches long shall be used for attaching the third layer of wallboard to the steel studs at 12 inches on center.

APPENDIX FIGURE 17

Gypsum wallboard protected structural steel columns with steel studs/screw attachment system.

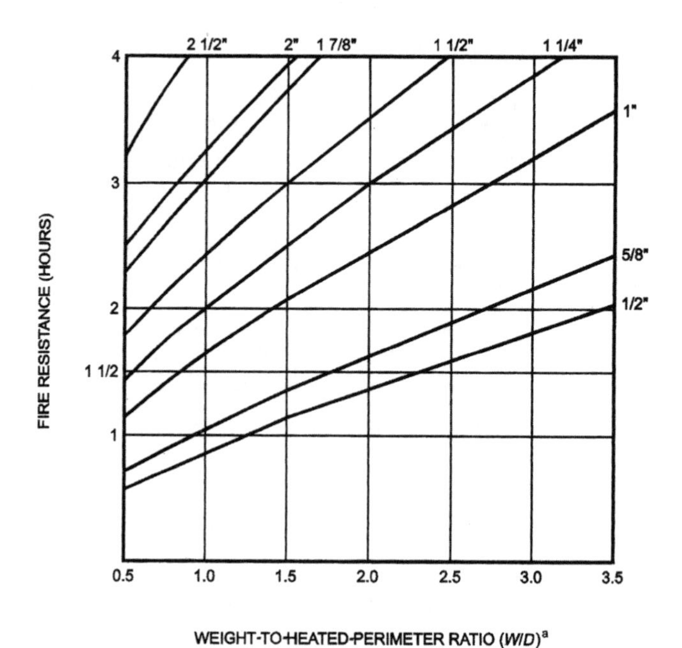

For SI: 1 inch = 25.4 mm, 1 pound per linear foot/inch = 0.059 kg/m/mm.

APPENDIX FIGURE 18

Fire resistance of structural steel columns protected with various thicknesses of type x gypsum wallboard.

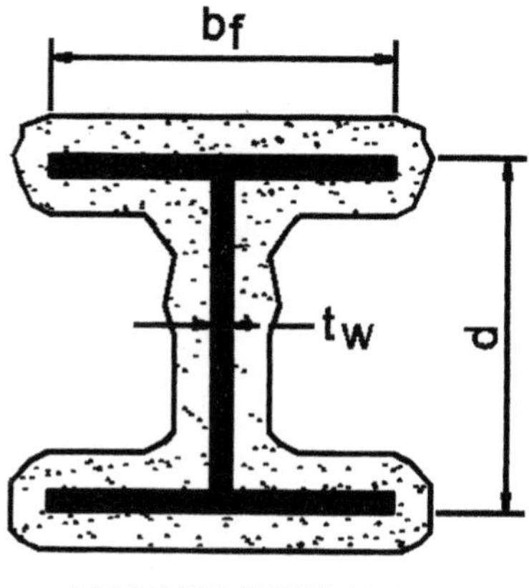

APPENDIX FIGURE 19

Wide flange structural steel columns with sprayed fire-resistant materials.

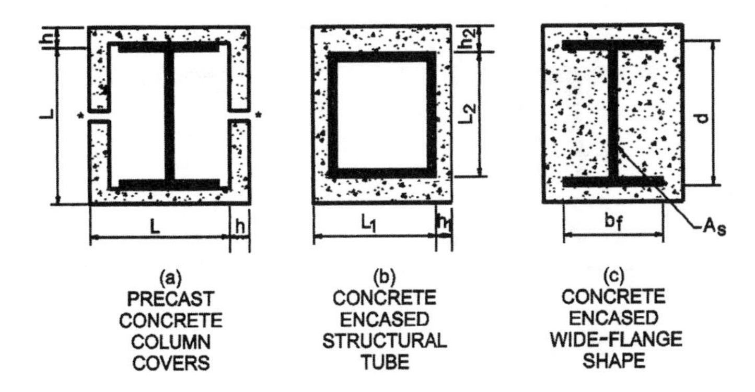

a. When the inside perimeter of the concrete protection is not square, L shall be taken as the average of L_1 and L_2. When the thickness of concrete cover is not constant, h shall be taken as the average of h_1 and h_2.

b. Joints shall be protected with a minimum 1 inch thickness of ceramic fiber blanket but in no case less than one-half the thickness of the column cover (see Section 720.2.1.3).

APPENDIX FIGURE 20

Concrete protected structural steel columns.

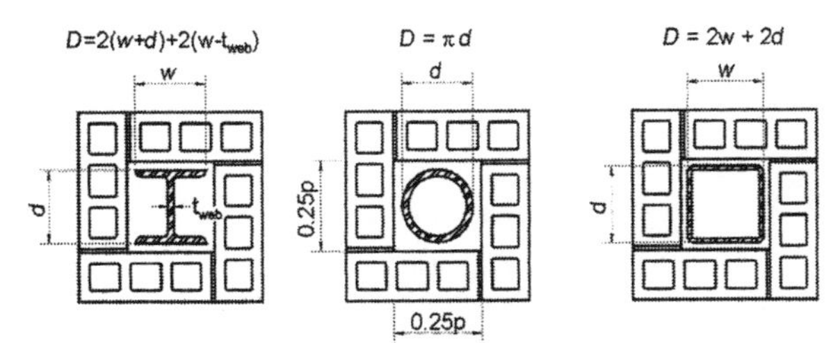

For SI: 1 inch = 25.4 mm.

d = Depth of a wide flange column, outside diameter of pipe column, or outside dimension of structural tubing column (inches).

t_{web} = Thickness of web of wide flange column (inches).

w = Width of flange of wide flange column (inches).

APPENDIX FIGURE 21

Concrete or clay masonry protected structural steel columns.

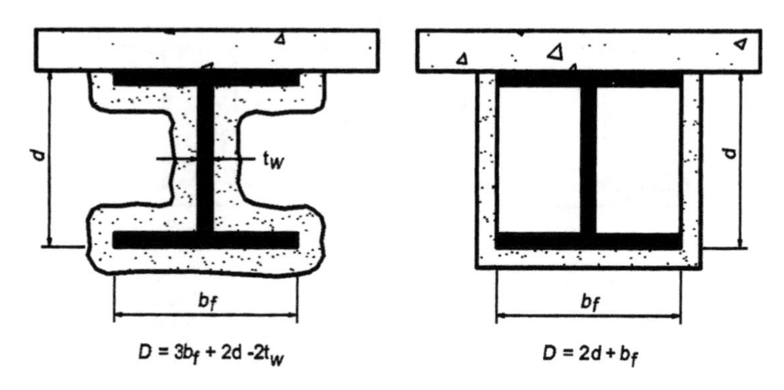

APPENDIX FIGURE 22

Determination of the heated perimeter of structural steel beams and grinder.

STRUCTURAL SHAPE	CONTOUR PROFILE	BOX PROFILE	STRUCTURAL SHAPE	CONTOUR PROFILE	BOX PROFILE
W14 × 233	2.49	3.65	W10 × 112	1.78	2.57
× 211	2.28	3.35	× 100	1.61	2.33
× 193	2.10	3.09	× 88	1.43	2.08
× 176	1.93	2.85	× 77	1.26	1.85
× 159	1.75	2.60	× 68	1.13	1.66
× 145	1.61	2.39	× 60	1.00	1.48
× 132	1.52	2.25	× 54	0.91	1.34
× 120	1.39	2.06	× 49	0.83	1.23
× 109	1.27	1.88	× 45	0.87	1.24
× 99	1.16	1.72	× 39	0.76	1.09
× 90	1.06	1.58	× 33	0.65	0.93
× 82	1.20	1.68			
× 74	1.09	1.53	W8 × 67	1.34	1.94
× 68	1.01	1.41	× 58	1.18	1.71
× 61	0.91	1.28	× 48	0.99S	1.44
× 53	0.89	1.21	× 40	0.83	1.23
× 48	0.81	1.10	× 35	0.73	1.08
× 43	0.73	0.99	× 31	0.65	0.97
			× 28	0.67	0.96
W12 × 190	2.46	3.51	× 24	0.58	0.83
× 170	2.22	3.20	× 21	0.57	0.77
× 152	2.01	2.90	× 18	0.49	0.67
× 136	1.82	2.63			
× 120	1.62	2.36	W6 × 25	0.69	1.00
× 106	1.44	2.11	× 20	0.56	0.82
× 96	1.32	1.93	× 16	0.57	0.78
× 87	1.20	1.76	× 15	0.42	0.63
× 79	1.10	1.61	× 12	0.43	0.60
× 72	1.00	1.48	× 9	0.33	0.46
× 65	0.91	1.35			
× 58	0.91	1.31	W5 × 19	0.64	0.93
× 53	0.84	1.20	× 16	0.54	0.80
× 50	0.89	1.23			
× 45	0.81	1.12	W4 × 13	0.54	0.79
× 40	0.72	1.00			

For SI: 1 pound per linear foot per inch = 0.059 kg/m/mm.

APPENDIX FIGURE 23

W/D ratios for steel columns.

Thermal conductivity (k_c)	0.95 Btu/hr · ft · °F	0.35 Btu/hr · ft · °F
Specific heat (c_c)	0.20 Btu/lb °F	0.20 Btu/lb °F
Density (P_c)	145 lb/ft^3	110 lb/ft^3
Equilibrium (free) moisture content (m) by volume	4%	5%

For SI: 1 inch = 25.4 mm, 1 foot = 304.8 mm, 1 lb/ft^3 = 16.0185 kg/m^3, Btu/hr · ft · °F = 1.731 W/(m · K).

APPENDIX FIGURE 24

Properties of concrete.

DENSITY (d_m) OF UNITS (lb/ft^3)	THERMAL CONDUCTIVITY (K) OF UNITS (Btu/hr · ft · °F)
Concrete Masonry Units	
80	0.207
85	0.228
90	0.252
95	0.278
100	0.308
105	0.340
110	0.376
115	0.416
120	0.459
125	0.508
130	0.561
135	0.620
140	0.685
145	0.758
150	0.837
Clay Masonry Units	
120	1.25
130	2.25

For SI: 1 pound per cubic foot = 16.0185 kg/m^3, Btu per hour · foot · °F = 1.731 W/(m · K).

APPENDIX FIGURE 25

Thermal conductivity of concrete or clay masonry units.

COLUMN SIZE	CLAY MASONRY DENSITY, POUNDS PER CUBIC FOOT	MINIMUM REQUIRED EQUIVALENT THICKNESS FOR FIRE-RESISTANCE RATING OF CLAY MASONRY PROTECTION ASSEMBLY, T_e (inches)			
		1-hour	2-hour	3-hour	4-hour
W14 × 82	120	1.23	2.42	3.41	4.29
	130	1.40	2.70	3.78	4.74
W14 × 68	120	1.34	2.54	3.54	4.43
	130	1.51	2.82	3.91	4.87
W14 × 53	120	1.43	2.65	3.65	4.54
	130	1.61	2.93	4.02	4.98
W14 × 43	120	1.54	2.76	3.77	4.66
	130	1.72	3.04	4.13	5.09
W12 × 72	120	1.32	2.52	3.51	4.40
	130	1.50	2.80	3.88	4.84
W12 × 58	120	1.40	2.61	3.61	4.50
	130	1.57	2.89	3.98	4.94
W12 × 50	120	1.43	2.65	3.66	4.55
	130	1.61	2.93	4.02	4.99
W12 × 40	120	1.54	2.77	3.78	4.67
	130	1.72	3.05	4.14	5.10

COLUMN SIZE	CLAY MASONRY DENSITY, POUNDS PER CUBIC FOOT	MINIMUM REQUIRED EQUIVALENT THICKNESS FOR FIRE-RESISTANCE RATING OF CLAY MASONRY PROTECTION ASSEMBLY, T_e (inches)			
		1-hour	2-hour	3-hour	4-hour
W10 × 68	120	1.27	2.46	3.26	4.35
	130	1.44	2.75	3.83	4.80
W10 × 54	120	1.40	2.61	3.62	4.51
	130	1.58	2.89	3.98	4.95
W10 × 45	120	1.44	2.66	3.67	4.57
	130	1.62	2.95	4.04	5.01
W10 × 33	120	1.59	2.82	3.84	4.73
	130	1.77	3.10	4.20	5.13
W8 × 40	120	1.47	2.70	3.71	4.61
	130	1.65	2.98	4.08	5.04
W8 × 31	120	1.59	2.82	3.84	4.73
	130	1.77	3.10	4.20	5.17
W8 × 24	120	1.66	2.90	3.92	4.82
	130	1.84	3.18	4.28	5.25
W8 × 18	120	1.75	3.00	4.01	4.91
	130	1.93	3.27	4.37	5.34

STEEL TUBING

NOMINAL TUBE SIZE (inches)	CLAY MASONRY DENSITY, POUNDS PER CUBIC FOOT	MINIMUM REQUIRED EQUIVALENT THICKNESS FOR FIRE-RESISTANCE RATING OF CLAY MASONRY PROTECTION ASSEMBLY, T_e (inches)			
		1-hour	2-hour	3-hour	4-hour
4 × 4 × 1/2 wall thickness	120	1.44	2.72	3.76	4.68
	130	1.62	3.00	4.12	5.11
4 × 4 × 3/8 wall thickness	120	1.56	2.84	3.88	4.78
	130	1.74	3.12	4.23	5.21
4 × 4 × 1/4 wall thickness	120	1.72	2.99	4.02	4.92
	130	1.89	3.26	4.37	5.34
6 × 6 × 1/2 wall thickness	120	1.33	2.58	3.62	4.52
	130	1.50	2.86	3.98	4.96
6 × 6 × 3/8 wall thickness	120	1.48	2.74	3.76	4.67
	130	1.65	3.01	4.13	5.10
6 × 6 × 1/4 wall thickness	120	1.66	2.91	3.94	4.84
	130	1.83	3.19	4.30	5.27
8 × 8 × 1/2 wall thickness	120	1.27	2.50	3.52	4.42
	130	1.44	2.78	3.89	4.86
8 × 8 × 3/8 wall thickness	120	1.43	2.67	3.69	4.59
	130	1.60	2.95	4.05	5.02
8 × 8 × 1/4 wall thickness	120	1.62	2.87	3.89	4.78
	130	1.79	3.14	4.24	5.21

STEEL PIPE

NOMINAL PIPE SIZE (inches)	CLAY MASONRY DENSITY, POUNDS PER CUBIC FOOT	MINIMUM REQUIRED EQUIVALENT THICKNESS FOR FIRE-RESISTANCE RATING OF CLAY MASONRY PROTECTION ASSEMBLY, T_e (inches)			
		1-hour	2-hour	3-hour	4-hour
4 double extra strong 0.674 wall thickness	120	1.26	2.55	3.60	4.52
	130	1.42	2.82	3.96	4.95
4 extra strong 0.337 wall thickness	120	1.60	2.89	3.92	4.83
	130	1.77	3.16	4.28	5.25
4 standard 0.237 wall thickness	120	1.74	3.02	4.05	4.95
	130	1.92	3.29	4.40	5.37
5 double extra strong 0.750 wall thickness	120	1.17	2.44	3.48	4.40
	130	1.33	2.72	3.84	4.83
5 extra strong 0.375 wall thickness	120	1.55	2.82	3.85	4.76
	130	1.72	3.09	4.21	5.18
5 standard 0.258 wall thickness	120	1.71	2.97	4.00	4.90
	130	1.88	3.24	4.35	5.32
6 double extra strong 0.864 wall thickness	120	1.04	2.28	3.32	4.23
	130	1.19	2.60	3.68	4.67
6 extra strong 0.432 wall thickness	120	1.45	2.71	3.75	4.65
	130	1.62	2.99	4.10	5.08
6 standard 0.280 wall thickness	120	1.65	2.91	3.94	4.84
	130	1.82	3.19	4.30	5.27

APPENDIX FIGURE 26

Fire resistance of clay masonry protected steel columns.

STRUCTURAL SHAPE	CONTOUR PROFILE	BOX PROFILE	STRUCTURAL SHAPE	CONTOUR PROFILE	BOX PROFILE
W36 × 300	2.47	3.33	× 68	0.92	1.21
× 280	2.31	3.12	× 62	0.92	1.14
× 260	2.16	2.92	× 55	0.82	1.02
× 245	2.04	2.76			
× 230	1.92	2.61	W21 × 147	1.83	2.60
× 210	1.94	2.45	× 132	1.66	2.35
× 194	1.80	2.28	× 122	1.54	2.19
× 182	1.69	2.15	× 111	1.41	2.01
× 170	1.59	2.01	× 101	1.29	1.84
× 160	1.50	1.90	× 93	1.38	1.80
× 150	1.41	1.79	× 83	1.24	1.62
× 135	1.28	1.63	× 73	1.10	1.44
			× 68	1.03	1.35
W33 × 241	2.11	2.86	× 62	0.94	1.23
× 221	1.94	2.64	× 57	0.93	1.17
× 201	1.78	2.42	× 50	0.83	1.04
× 152	1.51	1.94	× 44	0.73	0.92
× 141	1.41	1.80			
× 130	1.31	1.67	W18 × 119	1.69	2.42
× 118	1.19	1.53	× 106	1.52	2.18
			× 97	1.39	2.01
W30 × 211	2.00	2.74	× 86	1.24	1.80
× 191	1.82	2.50	× 76	1.11	1.60
× 173	1.66	2.28	× 71	1.21	1.59
× 132	1.45	1.85	× 65	1.11	1.47
× 124	1.37	1.75	× 60	1.03	1.36
× 116	1.28	1.65	× 55	0.95	1.26
× 108	1.20	1.54	× 50	0.87	1.15
× 99	1.10	1.42	× 46	0.86	1.09
			× 40	0.75	0.96
W27 × 178	1.85	2.55	× 35	0.66	0.85
× 161	1.68	2.33			
× 146	1.53	2.12	W16 × 100	1.56	2.25
× 114	1.36	1.76	× 89	1.40	2.03
× 102	1.23	1.59	× 77	1.22	1.78
× 94	1.13	1.47	× 67	1.07	1.56
× 84	1.02	1.33	× 57	1.07	1.43
			× 50	0.94	1.26
			× 45	0.85	1.15
W24 × 162	1.85	2.57	× 40	0.76	1.03
× 146	1.68	2.34	× 36	0.69	0.93
× 131	1.52	2.12	× 31	0.65	0.83
× 117	1.36	1.91	× 26	0.55	0.70
× 104	1.22	1.71			
× 94	1.26	1.63	W14 × 132	1.83	3.00
× 84	1.13	1.47	× 120	1.67	2.75
× 76	1.03	1.34	× 109	1.53	2.52

(continued)

APPENDIX FIGURE 27

Weight-to-heated-perimeter ratios (W/D) for typical wide flange beam and girder shapes.

STRUCTURAL SHAPE	CONTOUR PROFILE	BOX PROFILE	STRUCTURAL SHAPE	CONTOUR PROFILE	BOX PROFILE
× 99	1.39	2.31	× 30	0.79	1.12
× 90	1.27	2.11	× 26	0.69	0.98
× 82	1.41	2.12	× 22	0.59	0.84
× 74	1.28	1.93	× 19	0.59	0.78
× 68	1.19	1.78	× 17	0.54	0.70
× 61	1.07	1.61	× 15	0.48	0.63
× 53	1.03	1.48	× 12	0.38	0.51
× 48	0.94	1.35			
× 43	0.85	1.22	W8 × 67	1.61	2.55
× 38	0.79	1.09	× 58	1.41	2.26
× 34	0.71	0.98	× 48	1.18	1.91
× 30	0.63	0.87	× 40	1.00	1.63
× 26	0.61	0.79	× 35	0.88	1.44
× 22	0.52	0.68	× 31	0.79	1.29
			× 28	0.80	1.24
W12 × 87	1.44	2.34	× 24	0.69	1.07
× 79	1.32	2.14	× 21	0.66	0.96
× 72	1.20	1.97	× 18	0.57	0.84
× 65	1.09	1.79	× 15	0.54	0.74
× 58	1.08	1.69	× 13	0.47	0.65
× 53	0.99	1.55	× 10	0.37	0.51
× 50	1.04	1.54			
× 45	0.95	1.40	W6 × 25	0.82	1.33
× 40	0.85	1.25	× 20	0.67	1.09
× 35	0.79	1.11	× 16	0.66	0.96
× 30	0.69	0.96	× 15	0.51	0.83
× 26	0.60	0.84	× 12	0.51	0.75
× 22	0.61	0.77	× 9	0.39	0.57
× 19	0.53	0.67			
× 16	0.45	0.57	W5 × 19	0.76	1.24
× 14	0.40	0.50	× 16	0.65	1.07
W10 × 112	2.14	3.38	W4 × 13	0.65	1.05
× 100	1.93	3.07			
× 88	1.70	2.75			
× 77	1.52	2.45			
× 68	1.35	2.20			
× 60	1.20	1.97			
× 54	1.09	1.79			
× 49	0.99	1.64			
× 45	1.03	1.59			
× 39	0.94	1.40			
× 33	0.77	1.20			

For SI: Pounds per linear foot per inch = 0.059 kg/m/mm.

APPENDIX FIGURE 27

Weight-to-heated-perimeter ratios (W/D) for typical wide flange beam and girder shapes.

COLUMN SIZE	CLAY MASONRY DENSITY, POUNDS PER CUBIC FOOT	MINIMUM REQUIRED EQUIVALENT THICKNESS FOR FIRE-RESISTANCE RATING OF CLAY MASONRY PROTECTION ASSEMBLY, T_e (inches)			
		1-hour	2-hour	3-hour	4-hour
W14 × 82	120	1.23	2.42	3.41	4.29
	130	1.40	2.70	3.78	4.74
W14 × 68	120	1.34	2.54	3.54	4.43
	130	1.51	2.82	3.91	4.87
W14 × 53	120	1.43	2.65	3.65	4.54
	130	1.61	2.93	4.02	4.98
W14 × 43	120	1.54	2.76	3.77	4.66
	130	1.72	3.04	4.13	5.09
W12 × 72	120	1.32	2.52	3.51	4.40
	130	1.50	2.80	3.88	4.84
W12 × 58	120	1.40	2.61	3.61	4.50
	130	1.57	2.89	3.98	4.94
W12 × 50	120	1.43	2.65	3.66	4.55
	130	1.61	2.93	4.02	4.99
W12 × 40	120	1.54	2.77	3.78	4.67
	130	1.72	3.05	4.14	5.10

COLUMN SIZE	CLAY MASONRY DENSITY, POUNDS PER CUBIC FOOT	MINIMUM REQUIRED EQUIVALENT THICKNESS FOR FIRE-RESISTANCE RATING OF CLAY MASONRY PROTECTION ASSEMBLY, T_e (inches)			
		1-hour	2-hour	3-hour	4-hour
W10 × 68	120	1.27	2.46	3.26	4.35
	130	1.44	2.75	3.83	4.80
W10 × 54	120	1.40	2.61	3.62	4.51
	130	1.58	2.89	3.98	4.95
W10 × 45	120	1.44	2.66	3.67	4.57
	130	1.62	2.95	4.04	5.01
W10 × 33	120	1.59	2.82	3.84	4.73
	130	1.77	3.10	4.20	5.13
W8 × 40	120	1.47	2.70	3.71	4.61
	130	1.65	2.98	4.08	5.04
W8 × 31	120	1.59	2.82	3.84	4.73
	130	1.77	3.10	4.20	5.17
W8 × 24	120	1.66	2.90	3.92	4.82
	130	1.84	3.18	4.28	5.25
W8 × 18	120	1.75	3.00	4.01	4.91
	130	1.93	3.27	4.37	5.34

STEEL TUBING

NOMINAL TUBE SIZE (inches)	CLAY MASONRY DENSITY, POUNDS PER CUBIC FOOT	MINIMUM REQUIRED EQUIVALENT THICKNESS FOR FIRE-RESISTANCE RATING OF CLAY MASONRY PROTECTION ASSEMBLY, T_e (inches)			
		1-hour	2-hour	3-hour	4-hour
4 × 4 × $^1/_2$ wall thickness	120	1.44	2.72	3.76	4.68
	130	1.62	3.00	4.12	5.11
4 × 4 × $^3/_8$ wall thickness	120	1.56	2.84	3.88	4.78
	130	1.74	3.12	4.23	5.21
4 × 4 × $^1/_4$ wall thickness	120	1.72	2.99	4.02	4.92
	130	1.89	3.26	4.37	5.34
6 × 6 × $^1/_2$ wall thickness	120	1.33	2.58	3.62	4.52
	130	1.50	2.86	3.98	4.96
6 × 6 × $^3/_8$ wall thickness	120	1.48	2.74	3.76	4.67
	130	1.65	3.01	4.13	5.10
6 × 6 × $^1/_4$ wall thickness	120	1.66	2.91	3.94	4.84
	130	1.83	3.19	4.30	5.27
8 × 8 × $^1/_2$ wall thickness	120	1.27	2.50	3.52	4.42
	130	1.44	2.78	3.89	4.86
8 × 8 × $^3/_8$ wall thickness	120	1.43	2.67	3.69	4.59
	130	1.60	2.95	4.05	5.02
8 × 8 × $^1/_4$ wall thickness	120	1.62	2.87	3.89	4.78
	130	1.79	3.14	4.24	5.21

STEEL PIPE

NOMINAL PIPE SIZE (inches)	CLAY MASONRY DENSITY, POUNDS PER CUBIC FOOT	MINIMUM REQUIRED EQUIVALENT THICKNESS FOR FIRE-RESISTANCE RATING OF CLAY MASONRY PROTECTION ASSEMBLY, T_e (inches)			
		1-hour	2-hour	3-hour	4-hour
4 double extra strong 0.674 wall thickness	120	1.26	2.55	3.60	4.52
	130	1.42	2.82	3.96	4.95
4 extra strong 0.337 wall thickness	120	1.60	2.89	3.92	4.83
	130	1.77	3.16	4.28	5.25
4 standard 0.237 wall thickness	120	1.74	3.02	4.05	4.95
	130	1.92	3.29	4.40	5.37
5 double extra strong 0.750 wall thickness	120	1.17	2.44	3.48	4.40
	130	1.33	2.72	3.84	4.83
5 extra strong 0.375 wall thickness	120	1.55	2.82	3.85	4.76
	130	1.72	3.09	4.21	5.18
5 standard 0.258 wall thickness	120	1.71	2.97	4.00	4.90
	130	1.88	3.24	4.35	5.32
6 double extra strong 0.864 wall thickness	120	1.04	2.28	3.32	4.23
	130	1.19	2.60	3.68	4.67
6 extra strong 0.432 wall thickness	120	1.45	2.71	3.75	4.65
	130	1.62	2.99	4.10	5.08
6 standard 0.280 wall thickness	120	1.65	2.91	3.94	4.84
	130	1.82	3.19	4.30	5.27

APPENDIX FIGURE 28

Fire resistance of clay masonry protected steel columns.

<table>
<tr><th rowspan="2">STRUCTURAL SHAPE</th><th colspan="5">FIRE-RESISTANCE RATING (hours)</th></tr>
<tr><th>1</th><th>1 1/2</th><th>2</th><th>3</th><th>4</th></tr>
<tr><td>W14 × 233</td><td rowspan="7">1</td><td rowspan="5">1</td><td rowspan="3">1</td><td rowspan="2">1 1/2</td><td>2</td></tr>
<tr><td>× 176</td><td rowspan="3">2 1/2</td></tr>
<tr><td>× 132</td><td rowspan="3">2</td></tr>
<tr><td>× 90</td><td rowspan="4">1 1/2</td></tr>
<tr><td>× 61</td><td rowspan="3">3</td></tr>
<tr><td>× 48</td><td rowspan="2">1 1/2</td><td rowspan="2">2 1/2</td></tr>
<tr><td>× 43</td></tr>
<tr><td>W12 × 152</td><td rowspan="5">1</td><td rowspan="2">1</td><td rowspan="2">1</td><td rowspan="3">2</td><td rowspan="2">2 1/2</td></tr>
<tr><td>× 96</td></tr>
<tr><td>× 65</td><td rowspan="3">1 1/2</td><td rowspan="3">1 1/2</td><td rowspan="3">3</td></tr>
<tr><td>× 50</td><td rowspan="2">2 1/2</td></tr>
<tr><td>× 40</td></tr>
<tr><td>W10 × 88</td><td>1</td><td rowspan="5">1 1/2</td><td rowspan="4">1 1/2</td><td>2</td><td rowspan="3">3</td></tr>
<tr><td>× 49</td><td rowspan="4">1</td><td rowspan="4">2 1/2</td></tr>
<tr><td>× 45</td></tr>
<tr><td>× 39</td><td rowspan="2">3 1/2</td></tr>
<tr><td>× 33</td><td>2</td></tr>
<tr><td>W8 × 67</td><td rowspan="6">1</td><td rowspan="2">1</td><td rowspan="3">1 1/2</td><td rowspan="3">2 1/2</td><td rowspan="2">3</td></tr>
<tr><td>× 58</td></tr>
<tr><td>× 48</td><td rowspan="4">1 1/2</td><td rowspan="3">3 1/2</td></tr>
<tr><td>× 31</td><td rowspan="3">2</td><td rowspan="3">3</td></tr>
<tr><td>× 21</td></tr>
<tr><td>× 18</td><td>4</td></tr>
<tr><td>W6 × 25</td><td rowspan="3">1</td><td>1 1/2</td><td>2</td><td rowspan="4">3</td><td>3 1/2</td></tr>
<tr><td>× 20</td><td rowspan="4">2</td><td rowspan="4">2 1/2</td><td rowspan="4">4</td></tr>
<tr><td>× 16</td></tr>
<tr><td>× 15</td><td rowspan="2">1 1/2</td></tr>
<tr><td>× 9</td><td>3 1/2</td></tr>
</table>

For SI: 1 inch = 25.4 mm.

a. The tabulated thicknesses are based upon the assumed properties of normal-weight concrete given in Table 721.5.1(2).

APPENDIX FIGURE 29

Minimum cover (inch) for steel columns encased in normal-weight concrete.

<table>
<tr><th rowspan="2">STRUCTURAL SHAPE</th><th colspan="5">FIRE-RESISTANCE RATING (HOURS)</th></tr>
<tr><th>1</th><th>1 1/2</th><th>2</th><th>3</th><th>4</th></tr>
<tr><td>W14 × 233</td><td rowspan="5">1</td><td rowspan="5">1</td><td rowspan="4">1</td><td>1</td><td rowspan="2">1 1/2</td></tr>
<tr><td>× 193</td><td rowspan="3">1 1/2</td></tr>
<tr><td>× 74</td><td>2</td></tr>
<tr><td>× 61</td><td rowspan="2">2 1/2</td></tr>
<tr><td>× 43</td><td>1 1/2</td><td>2</td></tr>
<tr><td>W12 × 65</td><td rowspan="3">1</td><td rowspan="3">1</td><td rowspan="2">1</td><td>1 1/2</td><td>2</td></tr>
<tr><td>× 53</td><td rowspan="2">2</td><td rowspan="2">2 1/2</td></tr>
<tr><td>× 40</td><td>1 1/2</td></tr>
<tr><td>W10 × 112</td><td rowspan="4">1</td><td rowspan="4">1</td><td rowspan="3">1</td><td rowspan="2">1 1/2</td><td rowspan="2">2</td></tr>
<tr><td>× 88</td></tr>
<tr><td>× 60</td><td rowspan="2">2</td><td rowspan="2">2 1/2</td></tr>
<tr><td>× 33</td><td>1 1/2</td></tr>
<tr><td>W8 × 35</td><td rowspan="4">1</td><td rowspan="3">1</td><td rowspan="4">1 1/2</td><td rowspan="2">2</td><td>2 1/2</td></tr>
<tr><td>× 28</td><td rowspan="3">3</td></tr>
<tr><td>× 24</td><td rowspan="2">2 1/2</td></tr>
<tr><td>× 18</td><td>1 1/2</td></tr>
</table>

For SI: 1 inch = 25.4 mm.

a. The tabulated thicknesses are based upon the assumed properties of structural lightweight concrete given in Table 721.5.1(2).

APPENDIX FIGURE 30

Minimum cover (inch) for steel columns encased in normal-weight concrete.

Index

Accessible entrances, 163
Accessible means of egress, 130
Accessible route, 161
Aircraft-related occupancies, 42
Aisles, 150
Alarm systems, 120
Alarms, 111
Allowable load-bearing values of soils, 219
Allowable stress design, 239
Alternative systems, 113
Area limitations, 64
Area modifications, 68
Automatic sprinkler systems, 42, 107

Braced wall panels, 241
Business group B, 4

Ceiling panels, 86
Children's playgrounds, 22
Class I, 116
Combustible materials, 40
Combustible storage, 46
Commercial cooking, 114
Concealed spaces, 100
Construction documents, 196
Conventional light-frame construction, 240
Conveyors, 53
Covered mall buildings, 18

Dead loads, 202
Decay, 238
Definitions, 63, 106
Detection systems, 120
Door handles, 139
Doors, 93, 134
Draft stopping, 101
Dressing rooms, 41

Driven pile foundations, 228
Dwelling units, 166

Earthquake loads, 210
Egress illumination, 129
Engineered wood products, 241
Emergency escape, 156
Emergency power, 60
Equipment platforms, 67
Excavation, 216
Exhaust ventilation, 61
Exit access, 148
Exit access doorways, 150
Exit signs, 145
Exits, 125, 150
Exterior exit ramps, 155
Exterior structural members, 92
Exterior walls, 77

Fiberboard, 235
Fill, 216
Fire barriers, 84
Fire classification, 186
Fire dampers, 97
Fire-resistance joint systems, 89
Fire-resistance rating, 75
Fire-resistance rating of structural members, 90
Fire-resistance separations, 21
Fire-retardant-treated lumber, 236
Fire tests, 75
Fire walls, 81
Flashing, 176
Floor joists, 240
Floor loads, 208
Floor number signs, 153
Floors, 138
Foam plastic insulation, 245
Footings, 219
Foundation investigations, 214
Foundations, 219

Gas detection, 59
Gates, 134
General design requirements, 198
Grading, 216
Group A, 107
Group A-1, 2
Group A-2, 2
Group A-3, 3, 108
Group A-4, 3, 108
Group A-5, 4, 108
Group E, 5, 109
Group F, 6
Group F-1, 6, 109
Group F-2, 8
Group H, 8, 109
Group H-5, 55
Group I, 110
Group I-2, 29
Group I-4, 13
Group M, 13
Group R, 13
Group S, 15
Group S-1, 10, 111
Group U, 16
Guards, 147

Hangers, 43
Hazardous materials, 46, 50
Height, 65
Height limitations, 64
High-rise buildings, 25
Horizontal assemblies, 86

Horizontal exists, 154

I-3 different occupancies, 32
I-3 facilities, 11
Institutional group I, 10
Interior finish, 249
Interior trim, 249

Landings, 138
Light-frame construction, 240
Light-transmitting plastic glazing, 252
Light-transmitting plastic interior signs, 254
Light-transmitting plastic roof panels, 253
Light-transmitting plastic skylight glazing, 254
Light-transmitting plastic wall panels, 252
Light-transmitting plastics, 250
Live loads, 202
Load combinations, 200
Loose-fill insulation, 103

Materials, 175, 187
Medical facilities, 29
Mezzanines, 65
Micropiles, 229
Minimum quality, 234
Minimum standards, 234
Mixed occupancy, 53
Mixed use, 73
Monitoring, 111
Occupancy, 73
Occupancy groups, 50, 61
Occupant load, 128
Opening protectives, 93
Outside walls, 78

Parapet construction, 80
Parking facilities, 165
Passenger loading facilities, 165
Performance requirements, 173, 185
Pier foundations, 223, 230
Pile foundations, 223
Plastic signs, 24
Plastic veneer, 250
Platform construction, 41
Platform lifts, 171
Platforms, 37
Playgrounds, 22
Power-operated sliding doors, 34
Purlins, 241

Rain loads, 208
Ramps, 144, 155
Reroofing, 192
Rescue, 156
Roof coverings, 187
Roof insulation, 190
Roof vents, 40
Rooftop structures, 191

Security, 24
Signage, 171
Sleeping units, 166
Sliding doors, 34
Smoke barriers, 85
Smoke control, 121
Smoke damper, 98
Smoke partitions, 85

Smokeproof enclosures, 153
Snow loads, 205
Soil lateral loads, 207
Soils investigation, 214
Sound-insulating materials, 103
Special occupancies, 169
Sprinkler systems, 42
Stage doors, 38
Stages, 37
Stairways, 140, 155
Standpipe systems, 116, 119
Storage, 50
Termites, 238
Thermal-insulating materials, 103
Trusses, 235
Turnstiles, 134

Underground buildings, 27
Unlimited area buildings, 70

Veneers, 177
Vertical openings, 35
Vinyl siding, 181

Wind loads, 205
Window sills, 181